what is geography?

what is geography?

Alastair Bonnett

910
BON

Los Angeles | London | New Delhi
Singapore | Washington DC

First published 2008

Reprinted 2009, 2010

SAGE Publications Ltd
1 Oliver's Yard
55 City Road
London EC1Y 1SP

SAGE Publications Inc.
2455 Teller Road
Thousand Oaks, California 91320

SAGE Publications India Pvt Ltd
B 1/I 1 Mohan Cooperative Industrial Area
Mathura Road
New Delhi 110 044

SAGE Publications Asia-Pacific Pte Ltd
33 Pekin Street #02-01
Far East Square
Singapore 048763

Library of Congress Control Number: 2007941278

British Library Cataloguing in Publication data

A catalogue record for this book is available from the British Library

ISBN 978-1-4129-1868-8
ISBN 978-1-4129-1869-5 (pbk)

FSC
www.fsc.org
MIX
Paper from
responsible sources
FSC® C013604

Typeset by C&M Digitals (P) Ltd, Chennai, India
Printed and bound in Great Britain by CPI Antony Rowe
Printed on paper from sustainable resources

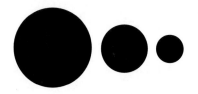

Contents

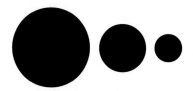

Introduction

We are all explorers. Even as tiny children we search out the limits of our world. A few years on, our imagination stretches further: fingers batting at a giddy plastic globe … a spinning top, gaudy with colour, representing perhaps the most ambitious idea possible, the world.

Geography is a fundamental fascination. It is also a core component of a good education. Yet a lot of people are not too sure what it is. They stumble over the question 'What is geography?' Perhaps they are worried by the scale and the implications of the obvious answer. For geography is about the world. To study geography is to study the world, both near and far.

This book introduces geography as one of humanity's big ideas. Geography is not just another academic specialism. Indeed, in an age when knowledge is fragmented into thousands of disciplines, geography can seem like a throwback. Its horizons are just too wide. After examining the way geology, climatology, ecology, environmental science and a number of human sciences evolved from geography, the historian Peter Bowler suggests that 'Geography is a classic example of a subject that can disappear as a separate entity, each of its functions siphoned off by a new specialisation'.[1]

But geography has not disappeared. Over the last 100 years its death has been predicted, even announced, many times. But it refuses to die. It seems that the desire for splintering the world into a kaleidoscope of intellectual shards also has its limits. More than this: that certain forms of knowledge familiar to people thousands of years ago (for geography is no spring chicken) still mean something to people today. Indeed, the need for world knowledge – for environmental knowledge, for international knowledge, for knowledge of the places and people beyond the parochial patch called home – could hardly be more pressing, more contemporary.

Figure 1 shows geography's area of study and its symbol. The world is geography's logo. It rolls around the letterheads of countless geographical societies, magazines and departmental websites. Admittedly, it is not usually as pretty as the world seen in this photograph. This image was photographed at a distance of 45,000 kilometres from earth by an astronaut aboard Apollo 17 on 7th December 1972. Soon named 'the Blue Marble', this is a picture at once beautiful and impressive yet disconcerting in its pocketsize neatness. Set against that inky, infinite surround, we are reminded of the fragility and unity of our world. NASA archivist Mike Gentry calls it 'the most widely distributed image in human history'.[2] It is the first and still the only time the whole earth had been photographed by a human eyewitness, for we have never since travelled such a distance.

Our world. It is an idea that provokes another: that our personal histories only make sense against the backdrop of six billion other personal histories, that our fates are intertwined. The Blue Marble is a portrait of a modern, interdependent, geographical consciousness.

The phrase 'personal *histories*' reminds us that geography has a twin. History and geography have much in common. Both are ancient but also contemporary. Both address seemingly limitless territory yet remain lodged in our imaginations; hard to grasp but indispensable. Immanuel Kant identified geography and history as the two basic forms of human knowledge, the one addressing things and events in space, the other things and events in time; the one reaching out, the other drilling down.[3] History, like geography, seems to be all-inclusive, endless in depth and range. It is intellectually omnivorous. But this does not allow us to dismiss history as incoherent. We know that history is about the past and we know that the past matters. In *What is History?* E. H. Carr argued that people living in different times have thought about the utility and nature of history in different ways. 'When we attempt to answer the question "What is history?"', Carr explained, 'our answer, consciously or unconsciously, reflects our own position in time'.[4] Those with a geographical bent of mind will want to add, 'and our position in the world'. But this is only to broaden Carr's point, which is to insist that, even though history is universally

Figure 1 View of the Earth as seen by the Apollo 17 crew traveling toward the moon. This translunar coast photograph extends from the Mediterranean Sea area to the Antarctica south polar ice cap. Original caption.
Courtesy of NASA

understood to be about the past, what people mean when they say 'history' is enormously variable.

Something similar can be said of geography. Across thousands of years and in many different communities world knowledge has been sought and created. But at particular times and in particular places, this project has taken on a particular shape and has been expressed in a particular language. For much of human history what we now call 'geographical knowledge' was determined by the demands of human survival, of the maintenance of bare life. Information on the immediate landscape, as well as of what might surround it, may have been meagre but it was essential. With the development of more complex societies geographical consciousness became more elaborate. And a recurrent theme began to emerge, namely that the world has a centre (unsurprisingly, this 'centre' was often the same place where the 'geography' was being imagined) and a more dangerous, somewhat strange, though perhaps enticingly exotic, periphery. It is not an unfamiliar model today. If we accept that the last few hundred years have been witness to the 'Westernisation of the world',[5] then we are also likely to see the planet in terms of centres and peripheries.

Industrial modernity shaped geography in its own image. The kind of questions increasingly associated with geography reflected this dominance. These questions turned on two interconnected themes: environmental and international change. More specifically, it is towards geography that people have turned when seeking answers to the questions, 'How and why has the environment altered?' and 'How and why do nations differ?' These questions, transformed into images, are postered across school rooms the world over. They are also well represented within television schedules and in the print media. The modern geographical agenda thrives on global diversity, on a boldly asserted cosmopolitanism. But it also asserts 'challenges' and 'problems' as central to the geography student's vocabulary. Indeed, to contemporary ears, the words 'environmental' and 'international' can seem a little bare without that pervasive suffix, 'crisis'.

It seems that the desire and need for knowledge of the world is a basic human attribute, yet the content and form of these desires and needs are changeable. Something else that is changeable is geography's audience. For much of the time since Eratosthenes (275–194 BC) coined the Greek word 'geo-graphy', or 'earth writing', over 2,200 years ago, written geographies were produced for and by a small elite. Their authors assumed that ordinary people had narrower horizons; that their interests were local and insular. Today, huge swaths of the earth's population have ready access to, and an apparent eagerness for information about places, peoples and events that are thousands

of miles from where they live. Many millions see international travel
and international awareness as normal parts of ordinary lives. Tales
of exotic destinations once had an aura of rarity and were often
preceded by a low bow from a diplomat returning to court. Today they
have become so common as to be banal. Stories of journeys to far-
flung destinations are told anywhere and by anyone. We are all, more
or less, plugged into our planet. Its availability and accessibility have
created a mass cosmopolitanism. Our wired-up, footloose, travel-
bugged world is stage to expanding and mutating forms of global
geographical awareness.

But as we imagine a world so easily spanned we also sense its vulnerability. Geography can still talk to us in the primal language of survival. To discuss environmental crisis may seem a distant, rather dry
exercise. But what is being discussed is survival. International knowledge can appear a globetrotter's luxury. But our era is one of world
wars and worldwide conflicts between opponents with the power to
destroy the planet. International knowledge too is about survival. Talk
of crisis is often overblown. But no other era has experienced the kind
of pressures on the environment that we have been witness to over the
past century. And in no other period have ordinary people become con-
scious of global conflict in the way we have today. It is a unique and
terrible kind of wisdom: we know we can destroy the world.

Why *What is Geography?*

There are many introductions to geography. Lots of academic
overviews, lots of school textbooks. For the most part they offer
summations of recent scholarship by academic geographers (if they
are university books) or recent key topics in environmental or international change (if they are school textbooks). What they do not tell
us is what geography is. To do that you have to step back and look
at the bigger picture. Geography has an interesting institutional
history (see Chapter 5). But if its story is confined to only one institutional form then it appears both random and disconnected. This
much is obvious if one looks at contemporary academic geography.

It covers everything from queer theory to quaternary science. Merely collating this vast body of activity will never lead to a plausible explanation of what geography is.

And we do need to know what geography is. Students do, journalists do, politicians do. Because they do not know. And because many of us have the nagging sense that it matters, that our world needs *this kind* of knowledge. I've taught geography in the university sector for many years now. Again and again students have asked me 'So, what is geography?'. Colleagues too, more wearily. For years I've batted the question away. 'Whatever geographers make it' was my glib response. To be honest even asking the question seemed vaguely wrong: like calling the police round to spoil the atmosphere of a care-free, interdisciplinary party. But as the issues that are so central to geography – issues of environmental and global crisis – have become ever more pressing, the luxury of evasion has become harder to afford. In an era in which geographical questions are the central questions of the day we need to know what geography is.

This little book is for anyone who wants an answer to the question posed in its title. *It is a personal statement.*[6] And it is full of contentious arguments. It explains what I think geography is. But I like to think it is more than that. This book recognises geography as a characteristically human enterprise. Geography is an attempt to find and impose order on a seemingly chaotic world; an attempt that is simultaneously modern and pre-modern, ancient and contemporary. The following chapters introduce us to this extraordinary ambition. Geography spans both the human and the natural sciences (Chapters 1 and 2); its obsessions mirror our urbanising, mobile world (Chapter 3); and its methods reflect the challenges of acquiring environmental and international knowledge (Chapter 4). Finally, Chapter 5 shows that geography has many institutional forms but, somehow, constantly escapes and defies them. No institutional cage is quite big enough for a desire to know the world.

To Know the World: Order and Power

Introduction

For most of human history the ability to interpret and represent the surrounding landscape has been an urgent need and an absolute necessity. Homo sapiens have also shown a consistent desire to order their world, to find *meaning* in it.

The oldest literatures we have are geographies. The founding texts in what, over 2000 years later, came to be called 'the Western canon', are about heroes of exploration. Ulysses's ten-year trip home to Ithaca (in Homer's *The Odyssey*, compiled or written from the ninth century BC) and the – even older – tale of Jason's search for the Golden Fleece, are part of a continuous tradition of expressing culture through narratives of voyages to distant places.

Much of the allure of Ulysses and Jason's epic journeys relies on their encounters with fantastic creatures, such as Centaurs and the Cyclops. The thrill of the outlandish tells us that geography is rarely a simple matter of bare facts, of dusty lists. It is an imaginative leap, an attempt to intellectually claim territory far beyond one's immediate grasp. Indeed, it appears that the more complex and settled societies become, the more they wish to reach beyond the confines of the familiar. The Greek term, *Okeanoio*, or Ocean, was given to the unimaginably large river that ancient Mediterranean civilisations thought circled the inhabited world (i.e., Europe, Asia and Africa). Where civilisations know geographical limits they also know the desire to cross them. *Ultima Thule*, a semi-mythical island (perhaps

Iceland or Shetland) glimpsed in the course of circumnavigations of Britain, was an object of yearning *because* it was beyond the boundaries of the known.[1] Lucius Seneca, who documented many aspects of Roman society, offered just such a dissatisfied, restless geography, whose emotional charge vibrates as strongly today as when he first expressed it in the first century AD.

> An age shall come, in later years,
> when Ocean shall loose creation's bonds,
> when the great planet shall stand revealed,
> and Tethys shall disclose new worlds,
> nor shall Thule be last among lands.[2]

Today, dreams of 'new worlds' are even more pervasive. Our world is far more known and rigidly organised than Seneca's. The thirst for exploration, for *terra incognito*, has turned into a generalised anxiety. The modern multitudes of roaming backpackers and weekend trippers search out new *Ultima Thules* of personal fulfilment and intimate discovery. And our imaginations are cast outwards as well as inwards: ever further into space, to parallel universes, into the delirious, free geographies of virtual reality.

This chapter tells the story of geography's mission to know the world. In both Chapters 1 and 2 we are following a similar intellectual journey. In each we find the same two basic and interrelated tendencies: first, to make order from the world; and second, to grasp the world as a complex whole. These ambitions are at once practical and abstract. They help us to sustain life but also to make sense of it. In Chapter 2 I address how the world came to be understood as a creation of Nature. Chapter 1 is focused upon how the world came to be seen as the home of women and men. It is a story with ancient roots but which confronts us with the modern dilemmas of ethnocentrism and global domination.

Ordering the World

When we look at the different ways people have found meaning within the world we address traditions and practices that vary from one society and from one period to another. However, there are some geographical ideas that appear universal.[3] I turn now to three such ideas: the search for a world story; the need for orientation; and the notion of centre and periphery.

Stories of the World

The desire to find order in the world is a fundamental human need. Traditionally, this impulse has been expressed through religious and spiritual beliefs. Indeed, perhaps the main function of religion is to make sense of the existence of the world. The first words of the Bible tell us that 'In the beginning God created the heaven and the earth'. It is a religious message but also a necessary explanation: the earth is an expression of divine will.

> *The heavens declare the glory of God; and the firmament sheweth his handywork.*[4]

With this information we can make sense of the seasons, the flow of the rivers, the distribution of animals and plants. The place of people in creation is also explained. In *Genesis* we read: 'let them have dominion over the fish of the sea, and over the fowl of the air, and over cattle, and over the earth, and over every creeping thing that creepth the earth'.[5] The world was made by God for us. Christianity, Islam and Judaism all share this faith. In many other traditions the act of creation is offered in even more detail: the deity literally moulds the earth, or the world is part and parcel of the creator's body. In Chinese mythology, the originator is Pan Gu. He wakes, dispelling chaos, and forming the earth: his breath becomes the wind and clouds, his body the mountains, his blood the waters, his skin and body hair the flowers and trees.

Yet it seems that people have rarely been *entirely* satisfied by such explanations. Even within the most religious of eras, a searching, challenging diversity of geographical questions has been asked and answered: questions about how and why the climate varies between different places; questions about how and why the world's peoples differ. Such questions have demanded evidence and a less mystical way of finding order in the world. Thus, within medieval Europe, rational geographical questions were pursued, often through the evaluation of observation and data. World regional geographies, the most influential of which was the encyclopaedic *De imagine mundi* (circa 1100), contain much that is outlandish.[6] However, they also attempt to accurately depict the global mosaic. For example, the unknown author of *De imagine mundi* tells how the Caucus mountains divide the Northern from the Southern countries of Asia and relates the principal characteristics of these countries and their inhabitants. We recognise as fact some of the descriptions but others as matters of opinion or as purely fanciful. The placing and description of the Nile, of Egypt (which is said to lie in Asia) and its capital Babylon (Cairo) seem evidence based (if not necessarily accurate). But what are we to make of the description of the country north of the Caucuses 'where mares conceive through the wind alone and give birth to foals that live only three years'?[7] The *combination* of the plausible with the creative reflects a pre-modern expectation that geographical gaps could and should be 'filled in' in the absence of knowledge.

What John Kirtland Wright called the 'geographical lore' of medieval Europe represents a mixture of storytelling, faith and rational knowledge.[8] If we accept this then we have to move away from the notion that the earth's story was once told in *purely* mythical and religious terms before undergoing a secular revolution.[9] An interest in real world processes and evidence seems to be an irrepressible component of the geographical tradition. Religious explanations of our world have never quite sufficed. But nor do meaningless lists of facts. Just as we must question the purely religious character of medieval geography, we can also query the notion that Europe moved to a post-medieval geography through ridding itself of the desire for the 'big picture'.

The development of new, scientific, global geographies in the sixteenth and seventeenth centuries was not led by atheists. In most branches of geography it was initiated by those who saw in the world evidence of an overall divine plan. It is a position summed up by John Ray in his *The Word of God Manifested in the Works of the*

Creation (published in 1691). Ray argued that the scientific method offered holy revelation. He called for the study of nature in close detail and an end to the tradition of equating scientific knowledge to familiarity with ancient texts.

> *Let it not suffice us to be Book-learn'd, to read what others have written, and to take things upon Trust ... but let us ourselves examine Things as we have Opportunity, and converse with Nature.*[10]

To 'converse with Nature' remains the modern way. But does it show us any overall 'plan', any wider, larger explanation by which we may make sense of our planet? It seems that if one *wants to find* a master plan, a transcendental purpose, then one will probably find it. And as the modern era has unfolded it has become apparent that such stories do not necessarily take a religious form. To look at the world as determined by evolution, the spirit of 'Gaia', or even the spirit of 'progress' is also to claim a greater scheme, an overarching idea.

Thus in tracing geography's secularisation we are not necessarily following a shift away from the search for the 'big picture'. The notion that geography can be divided into two fields of enquiry, the regional and the general (or global), has allowed geographers to be highly empiricist and descriptive at the specific scale while making claims for an *overall* and *overarching* logic at the general scale. This was a central concern for Bartholomaus Keckerman (1572–1609), the Protestant theologian who is often credited as a founding father of modern geography. His division between *geographica generalis*, which addressed the earth globally (most especially its climatic and physical forms), and *geographica specialis*, which addressed particular regions, proved to be highly influential. Another German, Bernhardus Varenius (1622–1650), developed the distinction:

> *Geography itself falls into two parts: one general, the other special. The former considers the earth in general, explaining its various parts and general affections. The latter, that is, special geography, observing general rules, considers, in the case of individual regions, their site, divisions, boundaries and other matters worth knowing.*[11]

Keckerman's and Varenius's vision for geography rested on the emergence of European intellectual assertiveness. Europeans were beginning to claim the entire world as both available for domination and available to be studied, compared and contrasted in its various parts. 'The greatest part of Geography', Varenius wrote, 'is founded only upon the Experience and Observations of those who have described the several Countries'.[12]

Who has such 'Experience'? Or, to put it another way, who makes the 'Observations' and who is to be the 'Observed'? Over the last 400 years the desire to find a story, an overarching explanatory narrative, for the world, has been profoundly influenced by European colonial and neo-colonial mastery. The Biblical instruction that the progeny of Adam and Eve are to 'have dominion' became racialised. In his 1893 poem 'A Song of the English', Rudyard Kipling declared that God had 'smote for us a pathway to the ends of all the Earth'.[13] For Thomas Hodgkin in 1896, 'It was the mission of the Anglo-Saxon race to penetrate into every part of the world, and to help in the great work of civilisation'.[14] By the end of the nineteenth century many European intellectuals saw human life on the planet in terms of an evolutionary struggle. Benjamin Kidd, who interpreted Darwinian thought into Social Darwinism, saw in European dominion 'the cosmic order of things' and, hence, the application of the guiding principle that must rule all life, namely 'energy, enterprise and social efficiency'.[15] A global racial order was established, introducing a 'Eurocentric' way of imposing order and meaning on the world.

The late twentieth century saw the demise of overtly racist expressions of Eurocentrism. Yet the hubris captured in the titles of influential treatises such as *The Ideas that Conquered the World*,[16] *Why the West Won*[17] and *The Triumph of the West*,[18] is utterly contemporary. John Roberts concludes *The Triumph of the West* by arguing that the West's real and final conquest is seen in its absorption of the entire planet: i.e., in its development from a local force with the power to dominate others into the *global culture of the modern*. 'What seems to be clear', says Roberts, 'is that the story of western civilisation is now the story of mankind'.[19] The notion that 'mankind' has a story, with a structure and direction – a beginning, a middle and, perhaps, an end – reminds us that Western supremacism offers another overarching narrative for life on earth. The story of Western power is far more materially grounded than are religious or spiritual explanations. But it too supplies answers to the kind of questions that such traditions have usually dealt with: Where are we going? (towards or through Western modernity); Who are we? (we are developed or developing peoples); What is the point of human life? (to be civilised and to be materially satisfied).

The epic journey into modernity has been talked about in utopian language, as 'the end of history'. Francis Fukuyama, who has popularised the phrase, offers a story of Westernisation achieving a state of perfection. '[T]he present form of social and political organisation' he reassures us, 'is *completely satisfying* to human beings in their most essential characteristics'.[20] This blissful vision relies on a consumerist model of human needs and wants: the radiant perfection of shopping. In Benjamin Barber's more critical description of what he calls our 'McWorld', we find both an account of a Westernised world and the strange rebirth of the religious experience of ecstasy. In McWorld, Barber argues, we find the future painted,

> in shimmering pastels, a busy portrait of onrushing economic, technological, and ecological forces that demand integration and uniformity and that mesmerise peoples everywhere with fast music, fast computers, and fast food – MTV, Macintosh and McDonald's – pressing nations into one homogenous theme park.[21]

From God's 'firmament' to the 'shimmering pastels' of globalisation, the world has been explained and ordered through a succession of extraordinary stories. It seems that when it comes to finding an overarching meaning for the world, the apparent predominance of the secular and the rational in the modern era is deceptive. For political and scientific narratives (such as evolution and Westernisation) offer their own 'big picture' of the function and future of life on earth. Again and again we must tell the tale of the world, each time different, each time the same.

Orientation

The world is a material reality. Humans need to give meaning to this reality. People have traditionally used the earth's physical and human features – topography, climate and settlement – to place themselves. Many of our older place names betray such origins (for example, Vienna (*Wien*), comes from the Celtic *Vedunia*, meaning 'forest brook'). Our ability to locate ourselves in relation to physical and human features demands *orientation*. It is an interesting word. *Oriens* is Latin for east, derived from the verb, *oriri* 'to rise'. The word contains a clue to one of the universal material realities that shapes the geographical imagination, the arc of the sun, its rising in the east and setting in the west. The sun's daily journey provides us with our most basic sense of orientation. And it reminds us that the single most important physical feature we see on earth is not on earth at all.

Literally speaking, 'orientation' means to face the east. Medieval maps of Christendom were drawn with Jerusalem at the centre and top. In this way maps were made to face east, or orientated. For the same reason, some early maps of the island of Britain had Wales at the bottom and East Anglia at the top (as seen on Richard Gough's map, from 1360). There is a particular Christian reason why nations should turn towards Jerusalem. But Christian orientation drew from a much wider and older tradition of veneration for the rising sun. The positioning of buildings towards the east and south can be traced back to pagan structures, such as Stonehenge. Conversely, the fact that the West is the place where the sun goes down, has ensured its enduring association with death. The West has been seen as a site of life's ending and, by extension, of completion, maturity and mystery. In the late fourth century the Christian theologian, Bishop Severian of Gabala in Syria, explained why God had placed the Garden of Eden in the east,

> *in order to cause [man] to understand that, just as the light of heaven moves towards the West, so the human race hastens towards death.*[22]

Unsurprisingly, it is in China, a country with a vast amount of contiguous land to its West, that we find the oldest heritage of discovering and interpreting these far horizons. Accounts of 'Traditions Regarding Western Countries' became a regular part of dynastic histories from the fifth-century AD. From early accounts of explorations as far west as Syria through to depictions of the modern West one can identify a continuous tradition of Chinese commentary on the exotic lands of the setting sun. From these early accounts we find depictions of another Middle Kingdom where,

> *The people are tall, and upright in their dealings, like the Han [Chinese], but wear foreign dress. They call their country another 'Middle Kingdom' [Mediterranean?] ... west of these there is the Hei-shi [black or dark river] which is reported to be the western terminus of the world.*[23]

The notion of light and of life rising in the East and setting in the West appears to be a universal form of human orientation. However, even such basic ideas are subject to change as new geopolitical realities arise.

The rise to power of 'the West' is signified in a number of ways. One of the least obvious, because it is so taken for granted, is that the western rump of the Asian landmass has been awarded the status of being a continent, namely Europe. Another is that the West has come to signify the finished point of civilisation. The German philosopher Georg Hegel gave an influential fillip to this train of thought when, in the early nineteenth century, he outlined his geography of the dawn of modernity, what he called (in another reference to our veneration of the sun) 'the Enlightenment'. In a famous passage from *The Philosophy of History* (1822), Hegel explained that the 'History of the World travels from East to West, for Europe is absolutely the end of History'.[24]

The orientation of the world between East and West has developed into an important tool for ordering global affairs. In 1978 the Palestinian literary critic, Edward Said, diagnosed a Western ideology which he called 'orientalism'.[25] Orientalism causes the countries of the East to be seen as ill-formed, exotic landscapes that require Western intervention. Of course, many nations straddle East and West. Russia is the most prominent example. From the late eighteenth century, the Russian ruling classes oriented themselves towards the West, self-consciously claiming for Russia a European identity. The Russian geographer Vasily Tatischev's designation, in the 1730s, of the Urals and the Caucasus as, respectively, the eastern and southern termini of Europe is of interest, not merely because of its survival to the present day, but because it reflects the emergence of a clear desire amongst the Russian elite to establish Europe and Asia, West and East, as distinct political entities. 'Asianism' *(Aziatchina)* came to be associated with everything that was old and rotten, everything that needed to be ripped out of both Russia and her colonies. This was not an emotion restricted to aristocrats. It was expressed even more forcefully by Russian revolutionaries. Mixing images of political reaction with those of decay and infestation, Leon Trotsky looked forward to the development of a clean, new, Western civilisation. Speaking of the 1917 Russian revolution he explained that it 'means the final break of the people with Asianism, with the seventeenth century, with holy Russia, with ikons and cockroaches ... an assimilation of the whole people with civilisation'.[26]

The geography of modernity (capitalist and socialist) is a re-orientation: the sun may still rise in the East but Enlightenment comes from the West. We have come full circle. The sun's light has been replaced by a new type of light, of the Enlightened consciousness, of industrialised development, of the (electric) light of freedom that spreads its way across the globe into all its 'dark corners'. This orientation leads us onto another of the basic principles by which we give order to our world, the notion of centre and periphery.

Centres and Peripheries

Thus saith the Lord GOD; This is Jerusalem: I have set it in the midst of the nations and countries that are round her.[27]

The need to find order in the world appears to be inextricable from the hunger to see the world in terms of centre and periphery. The most primordial of such 'spatial hierarchies' may be our personal knowledge that everyone and everything is, to a lesser or greater extent, distant from our own body. This self-centred view of the world is overlaid for most of us by the knowledge that, in the eyes of others, we are outside the centre. People in Newcastle, where I live, have a highly developed sense that they are perceived by people in the south of England as being on the periphery (and they aren't wrong: in the south I have been asked how I can endure life 'beyond the tree line'). It is galling to be peripheral. But it can usually be tempered by claims of centrality that take place at other scales (after all, Newcastle is the capital of the North East region).

However, if we accept that the entire globe has undergone a profound modernisation and, hence, Westernisation, then it also seems plausible to say that we live in the most centralised era in human history. While ancient centres, such as Christendom, or imperial China and Rome, were surrounded by terrain outside their control and ken, today the Western cosmopolitan imagination claims the entire planet within its discourse of universal values and universal progress. Today's world consciousness has grown bigger and more sophisticated. Yet it carries a level of arrogance that is historically unique.

In explaining the world geographers have for centuries also been explaining the familiar and the exotic. And it is for this reason that when we read geography we always need to ask, 'Who is it written for?' For only by knowing the audience can we understand why some parts of the earth are treated with an acute eye for detail and accuracy and others with careless brush strokes. A basic rule of thumb is that the further away somewhere is the more casual and fanciful the description of it is likely to be. This helps explain the following tenth century Chinese description of the far West (modern day Turkey):

> *There are lambs which grow in the ground; the inhabitants wait till they are about to sprout, and then screen them off by building walls to prevent the beasts which are at large outside from eating them up. The navel of these lambs is connected with the ground; when it is cut the animal will die.*[28]

It is marvellous stuff. The mythological geographies of the distant past still have a strange power. They are heroically bizarre. Yet the wide-eyed curiosity about the world 'out there' that feeds them remains recognisable and sympathetic.

Geography's roving disposition means that it always carries a trace of the make believe. This is a source of inspiration but also a perpetual worry. Both modern and classical geographers have often sought to pre-empt criticism of their own accuracy. One of the most inventive geographies, Pliny the Elder's *Natural History* (77 AD), swarms with semi-human freaks and monstrous beasts from Africa and India. But Pliny evidently feels he has to defend himself against a cynical readership. He offers a clever if disingenuous excuse:

> *there are some things that I do not doubt will appear fantastic and unbelievable to many. For who ever believed in Ethiopians before seeing them? Or what thing is not miraculous when first it comes to our attention?*[29]

The sternly practical geographical accounts of Strabo (63 or 64 BC–AD 24) had already set a pattern of reviling the accuracy of competing accounts of the inhabited world. The writers who depict pygmies, headless men and dog-headed humans in the far-flung places of the world, says Strabo, are not real geographers:

> *For it is self-evident that they are weaving in myths intentionally, not through ignorance of the facts, but through an intentional invention of the impossible, to gratify the taste for the marvellous and the entertaining.*[30]

Strabo[31] is one of the most earnest of the early geographers. His 17 volume *Geography* is the kind of monument of erudition that may be claimed as a founding text. Yet, despite his censorious attitude to other people's flights of fancy, Strabo often spliced together earlier, entirely speculative geographies with more reliable sources. Indeed, he combined accurate reportage with far-fetched accounts of distant lands. Strabo's description of Ireland (which he saw as the western limit of the habitable earth) exemplifies just how prone he was to colourful guesswork in the absence of knowledge.

[Ireland] lies beyond Britain but ... is such a wretched place to live on account of the cold that the regions on beyond are regarded as uninhabitable ... they are man-eaters as well as heavy eaters, and since, further, they count it an honourable thing, when their fathers die, to devour them, and openly to have intercourse, not only with the other women, but also with their mothers and sisters.[32]

Strabo knew next to nothing about Ireland. What he did have were some firm prejudices nurtured in an intellectual world centred on the 'middle earth', or Mediterranean about the cold and miserable north. The peripheries Strabo depicted served to glorify the Roman imperial world view. Strabo was born into a wealthy family in Amasya in Anatolia, in present day Turkey. He was a staunch advocate of Roman rule in his homeland. His *Geography* was a contribution to Roman control over a fractious empire. 'The greater part of geography subserves the needs of states', he explained, adding 'that geography as a whole has a direct bearing upon the activities of commanders'.[33]

However, the relationship between centre and periphery is rarely simply a matter of domination. Strabo romanticised exotic people such as the Scythians of Central Asia. Such people, he felt, had a nobility of spirit that the spread of civilisation was making increasingly rare.

our mode of life has spread its change for the worse to almost all peoples, introducing amongst them luxury and sensual pleasures and, to satisfy these vices, base artifices that lead to innumerable acts of greed. So then, much wickedness of this sort has fallen on the barbarian peoples also, on the Nomads as well as the rest.[34]

Hence, the periphery becomes 'untouched' and authentic; a site of romantic critique of the centre. The modern era has intensified this paradoxical relationship. Modernisation seems unstoppable. Yet it produces a desire to 'respect' the 'natural', 'colourful' cultures of a shrinking, non-modern world. In the last century this allure occasionally turned the tables on the centre–periphery relationship. Artists and writers were at the forefront of a dissatisfied 'primitivism'; an intellectual current that contrasts the inhuman, materialistic West with the culturally rich and more human non-West:

we fight and make money and fill our heads with politics ... [the Indian] has been content to discover the soul and surrender himself to spontaneity.[35]

Wherever Western civilisation is dominant, all human contact has disappeared, except contact from which money can be made.[36]

In this way social criticism was turned into a geographical argument. A surrealist map of the world was produced in 1929 (Figure 1.1) and provides a nice example of how centres and peripheries can be rearranged. In this map Europe (aside from rebellious Ireland and Russia) becomes tiny while Oceania, Alaska and Labrador (associated with uncorrupted Native Americans by the surrealists) are hugely inflated. The rather famished portions of Africa indicate that by 1929 the recent fashion for things African had diminished the charm of the continent for the self-consciously outré young men of surrealism. The foreign objects and areas subject to their interest were both adaptable and replaceable.

From Strabo to surrealism primitivist empathy does not necessarily reveal a genuine interest or concern with distant societies. The exotic is a fantasy object that offers a momentary thrill of transgression: the frisson of association with the extraordinary. But the relationship of centre and periphery is not a question merely of cultural sensibilities. There are differences of power between Rome and Scythia, the West and the Orient, that render primitivist romances further proof of the ability of the centre to control the meaning of the periphery and, hence, the world.

Modern Geography: The World of Trade and Nations

Geography thrives on mobility and expansion. During the imperial phases of Roman and Chinese power we find an upsurge of interest in geography. Similarly, the rise of economic globalisation over the past 400 years has enabled a new awareness of the world as well as

Figure 1.1   The surrealist map of the world map, 1929, attributed to Yves Tanguy

Source: Variétés, June 1929

new forms of ethnocentrism. The story of the emergence of industrialised economies and trade between nation states arbitrated by international agreements is inseparable from the rise of modern geography. It is a remarkable transition. Well into the nineteenth century one could still find people in Europe for whom the immediate locale consumed their geographical imagination. Such peasant and folk communities were part of a fast disappearing pre-modern relationship to place. Today they are gone, not simply in the West but across the world. We are all, more or less, plugged into the world economy and world politics.

Addressing the Manchester Geographical Society in 1884, the journalist and explorer H. M. Stanley explained the importance of teaching geography to the young. He offered a telling metaphor:

> *the configurations of the world chart appear as clearly defined as though they were the outlines of a man's real estate – the world is only*

*a huge breeding farm, and the various parts round about the shores
are like so many stalls at a market place – and the people therein are
only so many vendors and buyers.*[37]

The world is grasped by Stanley as an arena of trade. He goes on to
encourage his audience to think of geography as an education in
commerce:

*you must teach it your youths, that when they arrive at manhood each
may know that beyond these islands there lie vast regions where they
also may carve out fortunes ... You must extend it among the mature
men, that by exhibition of it they may be led to reflect, if in some little
known part of this world there may not lie as rich markets as any now
so earnestly competed for.* [38]

In the modern era the map emerged as a supremely practical tool,
designed to service commercial and colonial ambitions. The map seen
in Figure 1.2 is taken from a popular Victorian textbook, Gill's
Student's Geography. Its content is dated: huge swaths of India are no
longer given over to opium cultivation. But it remains a recognisable
product of the modern geographical world-view, in which access routes,
markets and production dominate the way countries are represented.

We take it for granted today that geography will impart economic information: that it will tell us about tonnages shipped, key
exports and where raw materials are extracted. That most maps
massively magnify transport lines passes without comment. The
only form of information that rivals the status of economics is political. The modern era is the age of nations. To offer a representation
of the world means to offer a representation of state boundaries.
Indeed, the act of establishing national borders has taken on a primordial aspect. Everything – including flora and fauna – is readily
imagined in national terms. Geographies of, for example, Turkey
and India, are addressing very recent categories which are, moreover, internally diverse and have contested borders. Yet to think
nationally has become second nature. The flowers illustrated in
The Flora of Turkey[39] may not know it, nor the birds flapping
through the pages of *Birds of India*[40], but nature is now subject to
the state.

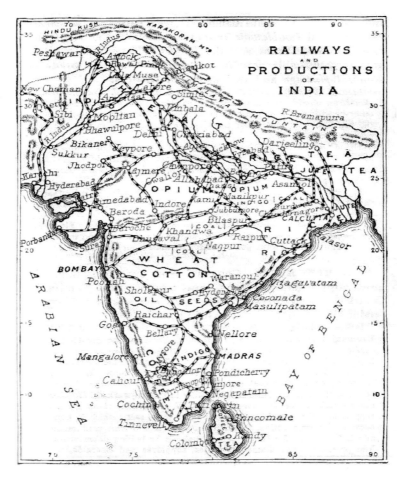

Figure 1.2   Railways and Productions of India, from Gill's *Student's Geography: Sixth Edition*

Source: Gill, 1903

The political map of the world is often placed alongside the physical map of the world. Both arise from the need for world information. However, since trade and war have been the defining forces of modern history, it is not surprising that the political map is usually to the fore. Newsrooms and many classrooms remain postered with 'The World: Political'. However, as we shall see in the next chapter, what we want and need to know about the world is changing. The rise of new concerns about the environment may soon diminish the priority given to nations and commerce in our maps of the world.

Knowing Others: Geography as the Modern Dilemma

Geography attempts to describe and explain the world and its peoples. There are many pitfalls in such an undertaking. One of the most fundamental is how we can know about people from other parts of the world. The increased mobility of people and information means that the issue of 'knowing others' has become a defining dilemma of the modern era. Prejudice, ignorance and stereotype are concerns within all intellectual fields of enquiry. But because of the nature of geography, because of its claim to produce information about other societies and other landscapes, it is forever shackled to this set of problems.

Contemporary geography agonises over anti-racism. Within primary and secondary education, many geography textbooks consciously strive to dismantle ethnic prejudice and expose students to human diversity. For some, geography's moral and political role is clear: it must seek to eliminate national enmities and enable global understanding. The political geographer, Isaiah Bowman, explained in *The New World* (1928), that 'we are now obliged to accept each other, for our modern world is shrinking and our mutual relations are made constantly more complex'.[41] Similarly, in his 1885 report to the Royal Geographical Society, the Russian anarchist Peter Kropotkin provided an influential statement of what today might be called geography's anti-racist function. Geography, he argued,

> *must teach us, from our earliest childhood, that we are all brethren, whatever our nationality ... geography must be – in so far as the school may do anything to counterbalance hostile influence – a means of*

dissipating these prejudices and of creating other feelings more worthy
of humanity.[42]

At the start of the twenty-first century such sentiments are so common that they have become something of a cliché. Yet geography's anti-racist remit is an ambivalent one. It often splices relativism with universalism: one moment saying 'respect difference', the next 'we are all the same'. It is a dilemma that can be found at work amongst some of the founders of the cosmopolitan tradition. We find the first recognisably modern concerns about 'relativism' in the essays of French nobleman Michel de Montaigne (1533–1592). Montaigne used the European encounter with the New World to challenge the notion that French manners and customs were superior. He was seeking to question the prevailing European view that the peoples and places of the Americas could and should be judged solely in European terms. The prevailing view was that the natives were barely human savages who had no title over their own land. Montaigne's position – as someone who implicitly claimed the capability of rising above such limitations – is that of the cosmopolitan, an individual of broader horizons than both the blinkered masses and the conservative elite. He provided a model of what we look for in the ideal geographer: a world-view that is unblinkered by prejudice and parochialism.

Montaigne's best known engagement with cultural prejudice came in his essay concerning Brazilian coastal peoples, 'On the cannibals'. His account was entirely drawn from secondary sources. Montaigne was not so much interested in providing a factual depiction of another culture, or the reality of cannibalism, as in asserting that even the most seemingly bizarre and exotic of social practices can be rationalised and justified if understood within their local context.

I find (from what has been told to me) that there is nothing savage or
barbarous about those peoples, but that every man calls barbarous
anything he is not accustomed to; it is indeed the case that we have no
other criterion of truth or right-reason than the example and form of
the opinions and customs of our own country. There we always find the
perfect religion, the perfect polity, the most developed and perfect way
of doing anything![43]

Montaigne is an early example of a European intellectual trying to
overturn prejudice. Yet his attempts to do so were beset with contradiction. The 'savages' of Brazil are to be respected he says, but
only because they conform to his notion of the Natural. Rousseau,
writing in 1755, as European colonialism seemed to be taking grip
of the earth, protested that,

> *In the two or three centuries since the inhabitants of Europe have been
> flooding into the other parts of the world, endlessly publishing new
> collections of voyages and travel, I am persuaded that we have come to
> know no other men except Europeans; moreover it appears from the
> ridiculous prejudices, which have not died out even among men of let-
> ters, that every author produces under the pompous name of the study
> of man nothing much more than the study of men of his own country.*[44]

The difficulties of producing knowledge about the world are com-
pounded by the fact that the West has been used to define the non-
West. The rest of the world is cast as 'developing', 'traditional' or
'uncorrupted'; all terms that imply a subservient status. In reaction,
over the last century, oppositional, alternative, world views arose
(which include ethnic projects like pan-Africanism and pan-Asianism,
as well as religious movements, notably radical Islamism), which have
sought to create counter-orthodoxies; other paths through (or away
from) modernity.

   The anger of such counter-movements has been heightened by the
fact that, in most countries, the spread and popularisation of a modern,
Western, geographical imagination has challenged native parochialism
and offered in its place, not a bestriding cosmopolitanism, but a galling
sense of marginality. In Japan the pre-eminent exponent of Westernisation, Fukuzawa Yukichi, issued a shocking challenge to insular
geographical traditions. His school textbook *World Geography* (1869)
placed Europe at the centre of world civilisation. Finding Japan
labelled as a 'primitive' country by Europeans, Fukuzawa's reaction
was to declare old Japanese geographies dead and call for a new and
assertive modern world consciousness:

> *While we think that we live on a flat, immobile earth, they know that
> it is round and in motion. While we regard Japan as the sacrosanct*

*islands of the gods, they have raced around the world, discovering new
lands and founding new nations.*[45]

Another, more ambivalent, challenge to 'old geography' can be found
in China. The traditional Sinocentric dismissal of foreign 'barbar-
ians' was challenged in the 1840s by geographers. Scholars such as
Wei Yuan and Xu Jiju began to introduce the public to the notion
that China was one country amongt many. However, this reform
was marked by contradiction.[46] Xu's world geography demoted
China: it was no longer the centre of creation. Yet his *Brief Account
of the Maritime Circuit* (1848) combines the new global consciousness with Sinocentrism, producing a novel and uneasy mix. He
opens his account with a statement on China's unparalleled importance. Xu is also keen to assert that Asia is the largest continent
and confirms an existing assumption that foreigners living in China
would come to resemble the Chinese (with, for example, their hair
becoming black). A similar ambivalence can be identified throughout the last century and into our own time. Students reading the
Chinese schoolbook *Geography of the World* (1942), written by Hu
Huanyong, learnt about the wide array of nations across the world.
Yet they also discovered that China has 'the longest history, the
highest culture, the largest population, a great and proud country'.[47]
In an era of racial and political struggle, they were also told that
China's position was reinforced by Japan and Turkey; 'two yellow
upcoming youngsters'. Together these nations would battle the
'whites', whose 'main camp' was in Europe.[48] Such bullishness contains a defensive quality. It reflects the erosion of certainty and its
replacement by self-conscious defiance against Western global
power.

Modern geography offers an assertion of who 'we' are in a world
of others. Yet it also undermines parochialism and, hence, 'our own'
distinctiveness. Across the world today classroom and popular geography books maintain a strange duality. Readers are told that their
nation is unique, special, and important. Geography is used to fill
the minds of the young with patriotic knowledge. Yet they are also
told that they live on a globalised planet and that geography is the
gateway to world knowledge. Over the past 100 years or so this

disjunction of ambitions has been most starkly drawn outside the West. For it is here that the need to use geography to affirm the nation has been most pressing. Within the West, by contrast, it has become increasingly common, especially within the larger nations, to identify a geographical education with a beneficent cosmopolitanism. Yet it is easy to be 'cosmopolitan' when the world is being shaped in your image. This cheerful certainty can make Western geography look complacent; a little too pleased with its willingness to 'include others'. The sunny disposition may not last.

Despite the vigorous nature of the Western claim, the world is not easily mastered. The rise of Asian power at the end of the twentieth century is provoking a new willingness to talk about 'multiple modernities'.[49] Western power may, finally, be on the wane. When it comes the oft-heralded decline of the West is likely to occasion the same mixture of adaptation and self-conscious defiance witnessed in other societies whose geographical imaginations have been overturned.

Conclusion

Homo sapiens are thinking animals. Thinking is the act of ordering experience into knowledge. Without order we are lost in a miasma of instinct and sensation. The most important expression of the desire to find order is the ordering of the world, both around the corner and beyond the horizon. This universal ambition gives us our geographical imagination. It is a universal but not an unchanging project. The modern era has seen the rearticulation and reshaping of very old concerns about human survival into a new language of international conflict and environmental crises. Geography is not a specialism, like sociology or geology. Nor is it the product only of recent history. Geography, like history, is an infuriating but vital combination of the modern and pre-modern. Its ambition is absurdly vast. But we know it would be more absurd to abandon it.

This chapter has introduced one of the two main ways geography attempts to find order in our world. It has looked at the world as a peopled place (in the next chapter I turn to the world as a natural place). One of modern geography's tasks is to make sense of this international and, perhaps, increasingly post-national, realm. We have seen how this ambition has emerged over the past few hundred years. Geography has become the international discipline, with an agenda that reflects the development of a global, commercial civilisation. We have also seen how geography's modern role exposes deep-seated dilemmas and ambiguities, problems that revolve around the imbalance of power between different parts of the world.

2 ●● To Know the World: People and Nature

Introduction

It is oddly easy to imagine: the view from space; the blue earth, loosely veiled with white cloud. Satellites take us closer. They begin to pick up the marks of human activity, clots of cities, vast irrigated plains, rivers channelled, straightened and damned. Almost everywhere the Earth's surface reveals the imprint of a numerous and highly organised species.

Our experience of the world is both of something made by humans and as something we call 'natural' (i.e., not made by humans). Because we are self-aware creatures we see ourselves as above nature or, at least, as acting upon it as if it was apart from us. Yet this is a troubling distinction. For in trying to describe nature we stumble into a dilemma: if the physical world is natural, does that make what we do to it *unnatural*? More questions follow: What should our relationship to nature be? Can we do what we like with nature? Should it be left alone?

But it is too late for that last query. We have already changed the earth; the flora and fauna are what we have left or care to keep; its landscapes have been dramatically altered by us and our pollutants have modified the soil and the climate.

Geography is the world discipline, and the world is both natural and human. Geography has sought to find order in both, and to study the relationship between the two. One of the great monographs of twentieth century geography is by Clarence Glacken and has the mysterious title *Traces on the Rhodian Shore* (1976).[1] The title refers to an ancient legend. The Greek philosopher Aristippus, shipwrecked and lost on an unknown island (the island of Rhodes), came across some geometric markings in the sand. 'Let us be of good hope' Aristippus cries out, 'for indeed I see the traces of men'. Glacken's survey of 2,300 years of environmental theories repeats this *cri de couer*: 'What is most striking in conceptions of nature, even mythical ones', he tells us, 'is the yearning for purpose and order'.[2]

The desire and ability to grasp the world intellectually is geography's defining act. The Greek scholar Eratosthenes, who gave us the word 'geography', provides a fine example. Not only is he credited with drawing one of the first world maps, he also devised a grid-based system to locate places on the earth, sketched the route of the Nile and worked out why it flooded (he argued it was due to seasonal rains near its source). Eratosthenes's quest to know the earth also prompted him to attempt to calculate its circumference. It may seem a prosaic statistic (we know today that it is 24,859 miles). But over 2000 years ago it represented a startling application of pure reason to a geographical problem. Eratosthenes knew that on the summer solstice at Syene (Aswan), the sun's rays left no shadow, falling directly down a well. Eratosthenes knew the distance from this well to Alexandria, the city where he worked as chief librarian, was 500 miles. He also knew that on the summer solstice in that city the sun's rays fell at an angle of 7 degrees. Since Eratosthenes already understood that the light from the sun travels in parallel lines and that the earth is round, he had enough data to calculate the circumference of the earth. He worked it out at 25,000 miles.[3]

There are two parts to this chapter. In the first I look at the relationship between people and the environment. In the second part I sketch a couple of theories of the world as a physical system. In the former the relationship between people and environment is to the fore. In the latter people are seemingly absent and the earth is represented as an arena of matter and force. However, both approaches are driven by the urge to turn the chaos of brute experience into order. They are both ways of providing meaning for the world.

People↔Nature

The relationship between people and nature has not always been framed by an assumption of crisis. Yet the insistence that neither nature nor humankind can be understood separately, that they are connected, has made geography uniquely sensitive to the idea that mistreating one may have adverse consequences for the other.

Discussion of the connections between people and the environment has often focused upon the question of the how the latter shapes the former. The notion that the environment moulds behaviour and character is an old one. It has been espoused and rebuked, refined and restated, for thousands of years. One of the earliest examples is Hippocrates's treatise, *Airs, Waters, Places* (400 BC). Hippocrates offers some stark correlations. Where a landscape is well-watered and mild, he says, one will find people who are 'fleshy, ill-articulated, moist, lazy, and generally cowardly'. But

> *where the land is bare, waterless, rough, oppressed by winter's storms and burnt by the sun, there you will see men who are hard, lean, well-articulated, well-braced, and hairy.*[4]

This is environmental determinism and it has long provoked controversy. Strabo seems to have found it rather risible. He pointed out that,

> *as regards the various arts and faculties and institutions of mankind, most of them, when once men have made a beginning, flourish in any latitude whatsoever and in certain instances even in spite of the latitude; so that some local characteristics of a people come by nature, others by training and habit. For instance, it was not by nature that the Athenians were fond of letters, whereas the Lacedaemonians, and also the Thebans, who are still closer to the Athenians, were not so; but rather by habit.*[5]

Environmental determinism has traditionally been focused upon climate. In the early twentieth century many geographers sought to move away from its cruder assumptions. Alfred Hettner and Lucian Febvre offered an alternative in what Febvre termed 'possibilism'.[6] 'There are no necessities, only possibilities' argued Febvre.[7] The possibilists' core conviction was that nature is not mandatory but permissive. The possibilists emphasised that human society consists of choices and actions and that nature does not determine any one path but offers instead a range of possible outcomes.

In recent years the determinist/possibilist debate has been reignited. Jared Diamond's *Guns, Germs and Steel* (1998) and Alfred Crosby's *Ecological Imperialism* (1986) both try and answer the question, 'Why did the West come to dominate the planet?'[8] To do so they consider a wide range of environmental factors, such as food production, continental location and incidence of disease. Moving between possibilism and determinism Diamond and Crosby explain social forces within the context of natural limits and natural resources. In his next book, *Collapse: How Societies Choose to Fail or Succeed* (2005), Diamond charted numerous rises and falls of ancient civilisations in terms of five factors: climate change, hostile neighbours, trade partners, environmental challenges and, finally, how a society deals with its environmental problems.[9] Although his material is historical, Diamond's message is a contemporary one: the most important reason societies collapse is because they fail to take environmental crises seriously.

Diamond and Crosby are part of a venerable geographical tradition of connecting the 'broad-patterns of history' to environmental factors.[10] The ambition of intellectual synthesis remains most firmly associated with the German explorer Alexander von Humbolt. In 1834 he set out to complete an integrated geography of everything.

I have the crazy notion to depict in a single work the entire material universe, all that we know of the phenomena of heaven and earth, from the nebulae of stars to the geography of mosses and granite rocks – and in a vivid style that will stimulate and elicit feeling. Every great and important idea in my writing should be registered side by side with facts. It should portray an epoch in the spiritual genesis of mankind – in the knowledge of nature ... My title is Cosmos.[11]

For many years and well into the last century, Humbolt's ambition to approach the human and natural world as one, interconnected and whole, was at the forefront of academic geography. The idea of interrelationship, noted Jean Brunhes, 'must dominate every complete study of geographical facts'.[12] The 'geography of the whole', wrote Paul Vidal de la Blache, 'is in truth the highest goal of geographic study'. Attaching the idea to regional study he contended that '[t]he dominant idea in all geographical progress is that of territorial unity'.[13] This integrative project was both a reflection of and a rebellion against the modern age. The scale of its ambition and the range of knowledge it relies upon are characteristically modern. Yet the attempt to overcome specialist demarcations and offer a unified portrait of the human and the natural offends the modern trend towards the strict policing of intellectual boundaries. Once again, we are confronted with geography's paradoxical relationship with modernity: it is both of and against our time.

By the mid-twentieth century, the desire to integrate the human and natural sciences was often overshadowed by the pursuit of academic specialisms. To have a narrow, tightly defined field of interest and knowledge became synonymous with having something serious to say. By the late twentieth century many university geographers had developed a deep suspicion of synthesis. In the words of Derek Gregory:

Claims [for synthesis] have been advanced many times in the past, as we all know, but more often as pious hopes or rueful excuses than as serious propositions ... it hasn't got very far because the natural and the social sciences keep pulling [geography] in different directions.[14]

However, Gregory's overview ignores the fact that the last 50 years have seen the development of influential environmental movements for whom holism and synthesis are central ideologies. What his statement reveals is not that synthesis is a hopeless ambition, forever roped to 'pious hopes', but that universities are structured upon intellectual division. The crises that have created environmentalism are not going to go away. Indeed, they are, in part, a consequence of the schism between humans and the natural landscape occasioned by the rise of fragmented, specialist rationalities. Holism and synthesis question not only 'we moderns' sense of superiority over non-modern times and places but also the way we have institutionalised knowledge.

The Land Ethic

As religious observance declined across much of the industrialising world, a veneration of nature took its place. In 1798 William Wordsworth captured this emerging mood:

> And I have felt
> A presence that disturbs me with the joy
> Of elevated thoughts, a sense sublime
> Of something far more deeply interfused,
> Whose dwelling is the light of setting suns,
> And the round ocean, and the living air,
> And the blue sky, and in the mind of man –
> ... Therefore am I still
> A lover if the meadows and the woods
> And mountains, and of all that we behold
> From this green earth.[15]

This is a deeply personal and romantic vision. At the same time it shows us how the deeply personal can be connected to the realm of Nature. In this modern vision, it is not God or the divine but sublime Nature that provides an inspiring order in which the self finds both repose and exhilaration. The American conservationist George Perkins Marsh provided a practical translation of this emerging sensibility in *Man and Nature* 1965[1864]. It is a pioneering investigation of the impact of human activity on the face of the land. At a time of unhindered settlement and industrial expansion across the USA, Marsh lit a warning beacon. He explained that the unmoderated exploitation of the landscape can have unintended consequences. Here, for example, he summarises some of the consequences of forest clearance:

> *With the disappearance of the forest, all is changed. At one season, the earth parts from its warmth by radiation to an open sky – receives, at another, an immoderate heat from the unobstructed rays of the sun. Hence, the climate becomes excessive and the soil is alternately parched by the fevers of summer, and seared by the rigors of winter. Bleak winds sweep unresisted over its surface ... the melting snows and vernal rains, no longer absorbed by a loose and bibulous vegetable mould, rush over the frozen surface, and pour down the valleys seaward, instead of filling a retentive bed of absorbent earth, and storing up a supply of moisture to feed perennial springs.[16]*

Marsh attempted to ignite a sense of responsibility for, and long-term involvement with, the land. Chopping down a forest supplies an immediate need. But it has other, longer term, results; such as floods, soil erosion and the degradation of what was once fertile land. The decline of flora and fauna and the increase in dust particles and carbon dioxide in the atmosphere are also likely to follow.

Unfortunately, industrial modernity offered so much wealth to so many and so quickly that such early warnings, although they inspired a few and caught the attention of millions, acquired an image of impractical worthiness. Certainly, by the mid-twentieth century, when Aldo Leopold's *Sand County Almanac* 1966 [1949] voiced similar concerns, the scale of the problem had grown. Leopold argued for a return of what he called a 'land ethic', writing,

> *We abuse the land because we regard it as a commodity belonging to us. When we see land as a community to which we belong, we may begin to use it with love and respect.[17]*

The idea that the 'land ethic' had *disappeared* in the West posed the problem of how it was to be reintroduced. One solution has been to take inspiration from non-Western conservation practices. Figure 2.1 offers an itemisation of sustainable practices of modern-hunter gatherers collected by Brian Hayden.[18] The activities depicted by Hayden show a determination to look upon the use of natural resources as an indefinite process. Similarly, the contemporary revival of Confucianism and Daoism in East Asia is, in part, focused on the environmental precepts associated with these belief systems, such as the duty to avoid the contamination of land and water and the

Groups ranked by resource diversity	Resource diversity	Seasonality	Conservation Practices
Western Desert Aborigines	Very low	Pronounced	Consistent use of peripheral areas whenever possible; use of major waterholes last
Great Basin Numa	Very low	Pronounced	Dislike using fire-drives because brush requires many years to regenerate; antelope drives held only every 12 years to permit animals time to reproduce
Montagnais	Very low	Pronounced	Randomisation of kill areas via divination practices, preventing overexploitation
Aranda	Low–moderate	Pronounced	Sacred nature of major sites precluded hunting within 1-mile radius and provided game refuges in time of environmental stress, thus preventing overexploitation; totemic prohibitions may have also provided differential refuge areas for particular species; compassion for animals
Kung Bushmen	Moderate	Pronounced	Use of peripheral areas whenever rainfall permits; conservation of water by using seasonal sources rather than permanent ones when possible
Birhor	High	Moderate	Allow areas to lie 'fallow' for 1–4 years; prefer not to join groups in order to conserve limited game; conscious of needing conservation for long-term productivity
Mbuti	High	Negligible	Conscious effort to use every part of animal; never kill more than necessary for the day

Figure 2.1 Conservation practices and resource variability in some hunter-gatherers

Source: based on Hayden, 1981

avoidance of human–natural unbalance.[19] Reaching back to ancient traditions helps to create an aura of antiquity for environmentalism. Modern societies are fascinated by these disappearing remnants. But it is an enchantment that displays a kind of bad faith. It is pertinent to recall that it was only when they could be romantically cast as 'dying races' that native and tribal peoples became objects of exotic attraction. Similarly, 'indigenous knowledge' and 'traditional farming practices' are venerated and 'respected' today, even though (or because) their extinction is largely complete.

Nevertheless, towards the end of the twentieth century fears about human impact on the environment appeared to have moved from the margins to the mainstream of political debate. The argument that there are *Limits to Growth*[20] and that *Small is Beautiful*[21] began to be widely heard. Barry Commoner, author of *The Closing Circle* (1971), hailed the fact that '[t]he environment has been rediscovered by the people who live in it'.[22] Commoner pointed out that the 'sudden public concern with the environment'

> *has taken many people by surprise. After all, garbage, foul air, putrid water, and mindless noise are nothing new; the sights, smells, and sounds of pollution have become an accustomed burden of life.*[23]

What puzzles Commoner is that, although environmental degradation is far from new, a widespread awareness of it is very recent. The *popularisation* of environmental concern has been stimulated, it seems, by two factors. First, a growth in the number of people whose viewpoint on society looks beyond their own material wealth (that is, they hold what sociologists call 'post-materialist' values[24]). Second, an increase in tangible evidence of the damage being done to the planet. In 1962 Rachel Carson's *Silent Spring* drew the attention of a mass readership to the effects of pesticides on the food chain. Her message was a contemporary one but it also takes us back to the most ancient roots of geography. For Carson was connecting environmental knowledge to human survival.

> *How could intelligent beings seek to control a few unwanted species by a method that contaminated the entire environment and brought the threat of disease and death to their own kind?*[25]

As we enter the twenty-first century, it appears that environmentalism is widely accepted. There are few political parties that do not wish to associate themselves with green issues. And there is a colourful spectrum of green politics, from deep greens, who espouse the protection of nature 'for its own sake',[26] to lighter shades who point to the utility of ecological protection for economic growth. And yet the growth of environmentalism has had only a minor influence on the course of industrial development. In an era when the majority of the world's population live in cities, the 'land ethic' is an alluring but remote ideal, something beyond people's immediate experience. There have been some successes for environmentalism, especially in the reduction of air pollution in the West and the conservation of some species which would have otherwise become extinct (for example, many fish species). But when set against the decline of biodiversity and the growth of resource exploitation throughout the world these are minor victories.

We are caught in a dilemma. Our reverence towards Nature is a function of our alienation from nature. The further we are removed from it the more we romanticise it. We have become armchair 'lovers of nature', sentimentally but genuinely attached to something we have little experience of, and it seems, quite unable to think beyond our comfortable, industrialised way of life.

Planetary Overload

The spectre of environmental crisis is stalking the world. We are increasingly aware of nature not just as a local, immediate environment but as a global and interconnected system. Figure 2.2 provides a visual expression of the *recent* and *sudden* nature of major human incursions into the environment. Some of the chemical releases it charts are almost entirely contemporary. Even the slowest shifts, such as in deforestation and the reduction in terrestrial vertebrate diversity, can be seen to leap up from the late nineteenth century onwards as industrialisation and population growth began to kick-in on a larger, global, scale.

These tendencies are liable to change along with changes to human population size and settlement. Thus, for example, in China, rural

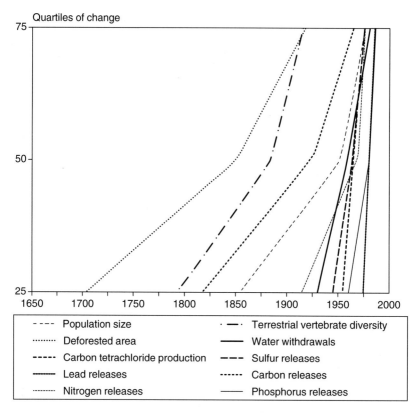

Figure 2.2 Trends in selected forms of human-induced transformation of environmental components
Source: Meyer and Turner III, 1995

migration to the cities, coupled with government tree planting schemes, mean that the past few years have witnessed reforestation.[27] The likely success of international initiatives to reduce pollution (the most influential so far being the Kyoto Protocol, 1997) is less clear. However, such counter-measures, most of which are very recent and difficult to gauge, are beyond the scope of this chapter. What concerns us here are the variety of ways the earth has been adversely affected by human activity. Below I offer some of the most important examples.

Climate Change

The greenhouse effect was first noted by Jean Fourier in 1824. He explained that the earth stays warm at night because the atmosphere traps air warmed by the sun. Without the greenhouse effect the earth would be about 30 degrees cooler and unable to support life. The principal 'greenhouse gas' is carbon dioxide. In 1894 Svante Arrhenius predicted that a doubling of carbon dioxide would lead to a rise in average temperatures of between 1.5 and 4 degrees. The twentieth century saw a 16-fold increase in the rate of emission of CO_2.[28] For much of the century this was not a source of worry. Indeed, Arrhenius looked forward to it: 'We may hope to enjoy ages with more equable and better climates, especially as regards the colder regions of the Earth'.[29] The meterologist Nils Ekholm reflected his era's technocratic confidence when he declared in 1901 that, 'it seems possible that Man will be able efficaciously to regulate the future climate of the Earth'.[30] Ekholm's confidence is explained, in part, by the fact that, until recently, it was assumed that climate change was an incredibly slow process. Only in the late twentieth century did the notion that temperature shifts can occur very quickly begin to take hold. Today, evidence of rapid changes in the past has become overwhelming. Indeed, ice cores indicate that fluctuations of 14 degrees have occurred in the distant past over a period of 10 years.[31]

Climate change provides the most prominent illustration of environmental *crisis*, of a new awareness that the relationship between people and nature can go wrong and can do so on a planetary scale. The consequence of human impact on the climate is disputed territory. Within some countries, most notably the USA, 'environmental sceptics' have found a ready audience, especially among those weary of environmentalists' penchant for doomsday scenarios.[32] How much the climate will heat up and how much of this heat is caused by human activity cannot be answered with certainty. However, predictive certainty is an impossible criterion. Looking at the distribution of warming across the world, as well as what we know of greenhouse gas production, the *likelihood* is that not only is the climate changing but that humans are partly to blame. The International Panel on Climate Change define 'likely' as 66–90 per cent probability and conclude that 'it is likely that anthropogenic [i.e., caused by humans] warming has had a discernible influence on many physical and biological systems'.[33] One of the important tasks that faces us now is to determine the *probable* effects of this process.

Since 1750 the concentration of carbon dioxide in the atmosphere has risen from 280 parts per million to 380 parts per million.[34] This change, which appears to have been caused largely by the burning of fossil fuels and deforestation, has been accompanied by a rise in the release of other greenhouse gases, notably nitrous oxide and methane. In total the effect of all these gases is equivalent to the concentration of 430 parts per million of carbon dioxide. This is a significant increase: it has been estimated that levels of these greenhouse gases in the environment today are at the highest point for 650,000 years.[35] Over the past 40 years, the oceans have soaked up 84 per cent of the newly generated warming effect of these greenhouse gases.[36] However, a warming trend also seems to be evident on land. Globally, the ten hottest years on record have all occurred since 1990.[37] An important prediction and warning was issued by the International Panel on Climate Change in 2001, who concluded that a warming of between 1.5–4.5 degrees could be expected to be produced by a doubling of carbon dioxide levels from the pre-industrial norm.[38] Such predictions are complicated by a number of feedback mechanisms. Warming is likely to lessen the ability of the earth's 'natural sinks' to absorb or retain carbon dioxide. Thus, for example, the ability of the oceans to absorb greenhouse gases is reduced by their warming; while the melting of frozen earth releases methane and carbon dioxide (methane emissions from Northern Siberia have increased by 60 per cent since the mid–1970s[39]).

The Stern Review (2006) has concluded that,

> *If annual emissions continued at today's levels, greenhouse gas levels would be close to double pre-industrial levels by the middle of the century. If this concentration were sustained temperatures are projected to eventually rise by 2–5C or even higher ... Near the middle of this range of warming (around 2–3C above today), the Earth would reach a temperature not seen since the middle Pliocene around 3 million years ago. This level of warming on a global scale is far outside the experience of human civilisation.[40]*

However, climate change has uneven effects across the globe. Shifts in weather patterns are likely to produce wetter, warmer conditions for higher latitudes and drier conditions, leading to desertification, for subtropical regions. A few of the negative consequences for people include declining crop yields in the subtropics, declining fish stock in acidified oceans (a result of increased carbon dioxide levels), flooding risks to low lying areas and the decay of ecosystems (with some 15–40 per cent of species facing extinction with two degrees warming).[41] It is estimated that as many as 200 million people could become 'environmental refugees' by mid century.[42]

Unless a global consensus forms and global action takes place to limit the release of greenhouse gases these are *likely* consequences. It is a daunting challenge for the geographical imagination. The voices calling for action range from Inuit fishermen to business people trading carbon emissions. It is an unprecedented moment. The outcome is uncertain. However, it is already clear that the threat of climate change offers a profound challenge to the way humanity understands its relationship with nature.

Air, Water and Soil

Dense human settlement, combined with industrial and domestic atmospheric pollutants, means that air quality in many cities around the world is poor. London's 'killer smog' of 1952, which killed several thousand people, was an early alert. Today, after years of pollution control and a shift towards service sector based economies, most Western cities have relatively clean air. By contrast, in the developing world the problem has become acute. Forest burning and industrial pollutants are responsible for permanent or seasonal pollution hazes that hang over much of South East Asia. Thus, for example, in 2006, the illegal burning of Indonesian forests created a haze that spread 2,250 miles, causing a choking pollution cloud that stretched from Malaysia far into the Pacific.[43] One study from 2002 estimated that every year nearly 800,000 deaths are the result of outdoor air pollution and that more than 1.6 million deaths are caused by the effect of indoor air pollution.[44] These statistics reflect the fact that it is nearly always the poorest parts of cities, where cheap, low quality fossil fuels are used for heat and cooking, where air pollution deaths are concentrated.

Fresh water makes up 3 per cent of the world's water. About 99 per cent of this is groundwater or ice. Running fresh water is a scarce resource. The scarcity and value of fresh water have been increased by the development of mass irrigation schemes that have lowered the level of fresh water lakes and damaged the sources of supply. The contraction of the Aral Sea provides a stark lesson in the overexploitation of a water resource. The Aral stretches either side of the border between Kazakhstan and Uzbekistan. The policy of the government of the USSR to expand crop production in Uzbekistan – to turn it into the cotton production region of the USSR – saw the development of massive irrigated plains alongside the Aral sea and its tributaries. The result was that the Aral lost two thirds of its surface area and 60 per cent of its water volume. This decline has been accompanied by a three-fold increase in salinity.[45] The exposed areas of the lake provide ideal conditions for the formation of dust storms. The storms whip up the dry soil and dump salty deposits over agricultural land, reducing its fertility. Today we find that people who live next to what was once the fourth largest fresh-water lake in the world have to have their water supply piped to them from elsewhere.

Soil degradation also appears to be linked to climate change. Thus, for example, the 2005 and 2006 Amazon droughts have been linked to the warming of the Atlantic and forest burning.[46] After drought the soil is nutritionally depleted, which threatens forest recovery. However, human activity is changing the nature of soil across the world in more direct ways. The two most important are through pollution (notably waste disposal and the knock-on effects of polluted air, such as 'acid rain') and the use of artificial additives. Soil productivity can be increased considerably by the application of a range of fertilisers and pesticides. These additives reduce or kill numerous forms of flora and fauna. They also change the composition of the soil itself. The resulting degradation of the microbial richness of the soil locks farmers into a reliance on ever greater use of pesticides and fertilisers. The most common artificial additives are nitrogen fertilisers. 'Human interference in the global nitrogen cycle is now at a far higher level than for either the carbon or sulphur cycles', argues Graham Harvey, a critic of the 'treadmill' of so-called 'chemical farming'.[47] At the same time it needs to be understood that 'chemical farming' is now an integral part of modern life. Valav Smil estimates that 40 per cent of the world's population are alive because of the increased fertility that has been made possible by nitrogen fertilisers.[48]

Humans and Other Species

Can there be too many people? If one looks at this question from the point of view of the environment the answer is yes. Species and habitat diversity have declined as the human population has expanded. A useful starting point for thinking about environmental impact is the following equation formulated by Paul Ehrlich and John Holdren:[49]

$$I = P \times A \times T$$

where:

I is the impact on the environment resulting from consumption
P is the population number
A is the consumption per capita (affluence)
T is the technology factor

i.e., environmental impact is the product of human population, multiplied by consumption per person, multiplied by the amount of resources needed to create things to be consumed.

It is a provocative equation. Yet more concrete information is, perhaps, even more telling. The United Nation's Millennium Ecosystem Assessment, which gathered data between 2001–2005 on ecosystem change, found that the 'rate of species extinction is several orders of magnitude higher than the natural or background rate'.[50] The report's authors go on to note that,

the most definite information, based on recorded extinctions of known species over the past 100 years, indicates extinction rates are around 100 times greater than rates characteristic of species in the fossil record. Other less direct estimates, some of which refer to extinctions

hundred of years into the future, estimate extinction rates 1,000 to
10,000 times higher than rates recorded among fossil lineages.[51]

Species diversity has declined as the human population has
expanded. One of the most striking facts about humans over the
past 200 years is their phenomenal numerical expansion. In 1800
there were 978,000 people in the world. Between 1800 to 1930
one billion were added to this figure. By 1960 there were three
billion people. According to the United Nations the world's population reached six billion on 12 October, 1999. The UN predicts the
population will reach 9.1 billion in 2050.[52]

Thomas Malthus, in *An Essay on the Principle of Population* (first
published in 1798), argued that 'the power of population is infinitely
greater than the power in the earth to produce subsistence for man'.[53]
In other words, that population can exceed a sustainable limit.
Malthus's ideas continue to provoke debate. In 1993, 59 national
scientific academies offered a joint statement calling for zero population growth.[54] However, not merely Malthusianism but any
attempt to identify population increase as a problem has been challenged, notably by social scientists. The idea that population growth
is manageable has found support from a number of studies of land
use change. Turner, Hanham and Portararo showed that the carrying capacity of land could be expanded by changes in agricultural
regime.[55] Ester Boserup has developed this line of argument and
reached the conclusion that by stimulating new patterns of land-use
a rising population might be a good thing.[56]

However, while offering a useful corrective to Malthusianism, a
focus on the political and technological possibility of human
increase relies on an increasingly untenable marginalisation of
environmental concerns. The past few decades have shown that,
irrespective of political regime, population growth has been accompanied by environmental degradation. Moreover, population increase
has only been made possible by the industrialisation of agriculture.
The most significant example of such intensification and innovation
is the so-called 'Green Revolution', which saw the introduction of
intensive pesticide and fertiliser use into farming across the majority of the world. Pest- and disease-resistant genetically modified

crops provide another example of a new technology, which can increase harvest crop quality and, hence, support a growing population. To return to Ehrlich and Holdren's formula, we might wish that a happy balance could be achieved between population growth and nature, with the effects of so many new humans ameliorated by the development of less materialist, less consumerist societies. However, no such trend is apparent. Population expansion across the majority of the world is certain to be accompanied by a significant rise in the consumption of natural resources. As the unsustainable levels of consumption characteristic of the West become a global norm, the problem of over-population is likely to become more pressing.

The Systems of Nature

Creation is the undesigned result of inevitable natural processes.
Democritus (470–380 BC)[57]

History is commonly understood to concern the *human* past. But geography is bigger still. Geography is not just about the earth as the home of humans. It is interested in the world *in toto*.

Geography has always been concerned with the natural processes that shape the earth. Geography's explorations of nature are also directed and driven by the very human need to find order in the world. Religious explanations of the earth's creation have been supplemented and, in the modern era, replaced, by the conviction that the environment may be understood as a set of natural systems bound by natural laws. This attitude has been accompanied by a new way of looking at the earth, as a 4.5 billion year old set of natural systems. Considering the planet in these terms may reinforce the kind of global consciousness we have already encountered in this book. But it also changes and challenges it. For it removes humans from centre stage. The arrogant belief that we are the point and pinnacle of the cosmos crumbles away, to be replaced by a recognition that humans are late arrivals, whose residence is not guaranteed and may be temporary.

Of all the natural systems that have fascinated observers, the study of climate has one of the longest histories. Climate research is also a clear example of how the geographical imagination, in trying to explain things local, is necessarily drawn into an examination of things global. Aristotle, in his *Meteorologica* (350 BC), connected the local with the global in order to offer a primitive model of the flow of winds across the known world.[58] The English scientist Edmond Halley, whose world map of trade winds was published in 1686, argued that solar heating causes air to rise and flow as winds. In the following century, George Hadley developed this idea, linking trade winds to the rotation of the earth. Hadley explained that hot air, rising at the tropics, cools and descends at higher latitudes, whereupon it flows back to the equator. In the late modern era climatology was advanced through developments in the chemistry and physics of the atmosphere. In 1878 Alfred Cornu established that short-wave ultraviolet solar radiation was being absorbed in the atmosphere. The absorber was discovered a couple of years later to be ozone. Another geochemical advance came with the publication of Svante Arrhenius's climate model in 1896, which showed how sensitive the earth's surface temperature is to the presence of carbon dioxide in the atmosphere.[59] Climate modelling reached a mathematical apogee in the first half of the twentieth century with Milutin Milankovitch's models of solar radiation, the variability of which he linked to changes in the earth's orbit, angle, and rotation.[60] By bringing astronomical data into climate modelling Milankovitch showed that to understand our planet's weather we must understand it in relation to other systems of nature, notably the solar system.

By the mid–late twentieth century the complexity of all these interconnected systems was leading many meteorologists to conclude that the weather is inherently unpredictable. In the early 1960s Edward Lorenz illustrated and developed the argument.[61] He did so by reference to the 'butterfly effect', i.e., that the beating of a single butterfly's wings might create tiny changes in the atmosphere that lead ultimately to the formation of a tornado. Lorenz's argument established what became known as 'chaos theory'. Chaos has proved a highly influential metaphor for discussing the limits of prediction across a wide variety of fields. It may seem an unruly intervention. After all, geographical enquiry is driven by a desire to find patterns, laws and order. In fact, chaos theory confirms this human imperative, for it seeks to develop what is known even to the point of establishing precisely what cannot (or is not yet) able to be determined.

However, the most fundamental shifts in our perception of the earth have not arisen from meteorology but from other branches of geography within the geo-sciences, such as geology and geophysics. The birth of modern geology represented a dramatic reconceptualisation of the world. Although the earth's rocks and landforms had been the subject of geographical description for thousands of years, the image of an immobile, *completed* earth had nearly always held sway. James Hutton (1726–97) shattered this image. His contribution was also early and astute enough to wrestle directly with the challenge it posed to alternative ways of finding order in the world. In his 'Theory of the Earth' (1788), Hutton takes us through different ways of looking at landscape geology in the company of the evidence that he had found of uplift, submersion and slow erosion. He challenges the Christian metaphor that compared the world to a perfect machine designed by an omniscient deity; a machine in which the earth's 'different parts are all adapted, in form, in quality, and in quantity, to a certain end'. Hutton wants us to abandon the notion that the earth can be understood as a finished, static artefact. Better he says to look at it as, 'an organised body ... [s]uch as has a constitution in which the necessary decay of the machine is naturally repaired'.[62]

Hutton is offering a model of nature not as a mechanism but as a self-regulating system; an active, interconnected set of continuous processes. This vision extinguished the stable, immobile earth and replaced it with a dynamic planet:

> *The formation of the present earth necessarily involves the destruction of the continents in the ancient world; and, by pursuing in our minds the natural operations of a former earth, we clearly see the origin of that land, by the fertility of which, we, and all the animated bodies of the sea, are fed. It is in like manner, that, contemplating the present operations of the globe, we may perceive the actual existence of those productive causes, which are now laying the foundation of land in the*

unfathomable regions of the sea, and which will, in time, give birth to future continents.[63]

The new geology replaced a human centred notion of time with a nature centred one which stretches out beyond the human imagination:

if the succession of worlds is established in the system of nature, it is in vain to look for any thing higher in the origin of the earth. The result, therefore, of our present enquiry is, that we find no vestige of a beginning, – no prospect of an end.[64]

For Hutton nature is order. But it is more than that, nature has wisdom. If left to itself, if seen across the numberless years, it provides a balanced, sustainable world and fit home for humans:

with such wisdom has nature ordered things in the economy of this world ... [that] in understanding the proper constitution of the present earth, we are led to know the source from whence had come all the materials which nature had employed in the construction of the world which appears.[65]

Twentieth century geophysics took Hutton's world vision and applied it to an analysis of how energy and matter shape the earth. The greatest change in our understanding of the world that these concerns produced was an awareness of the mobility of continents. The fit between South America and Africa had long been noticed. In 1800 Alexander von Humbolt speculated that they might once have been joined. This throwaway remark was elaborated into a theory of 'continental drift' by Alfred Wegener.[66] He presented his ideas to a sceptical audience at the Geological Association in Frankfurt on 6 January, 1912. His paper was called 'The Geophysical Basis for the Evolution of the Large-Scale Features of the Earth's Crust (Continents and Oceans)'. Wegener made the case for the movement of the earth's crust but he did not explain how and why it moved. It was only with the ability of researchers to chart the ocean bed in the 1950s that it became possible to move from a notion of 'continental drift' to one of plate tectonics. Harry Hess's investigation of the mid-Atlantic mountain ridge, which cuts down the middle of the Atlantic sea floor, showed how magma was pushing its way up to the surface, resulting in the sea floor spreading:

> *The continents do not plough through oceanic crust impelled by unknown forces, rather they ride passively on mantle material as it comes to the surface at the crest of the ridge and then moves laterally away from it.*[67]

Thus Wegener's vision of the earth was refined and given explanatory content. The earth's surface was now depicted as comprising vast plates, upon which are found both continents and oceans. There are some 30 of these plates, of various sizes, with a thickness that varies from five miles to 120 miles. The heat of the earth's core creates a viscous mantle across which the tectonic plates can judder and slide. Where the plates collide or diverge mountain ranges or ridges are formed. The Himalayas are being formed on the boundary between the Eurasian plate and the Indian/Australian plate, while the Andes are the site of collision of the Pacific and the South American plates. The Andes and the Himalayas are the products of a worldwide system of forces and movements.

The lives of Hutton, Wegener and Hess were separated by some 200 years but they each presented visions of the world that have profoundly shaped the way we look at our landscape. In some ways their work appears to reduce the importance and scale of the human occupation of the earth. Homo sapiens' few hundred thousand years on earth are small beer when set beside the aeons of geological time; our efforts at 'taming nature' are puny when set beside the forces that are moving continents beneath our feet. However, there may be another side to this manifest insignificance. For if we understand that the earth is not 'ours' but a domain upon which we rely, we may also develop a sense of vulnerability in relation to nature. The systems of nature may not 'need us', but through understanding them we gain a sobering appreciation of our species' likely impermanence.

Conclusion

On 26 December 2004 a huge earthquake (9.3 on the Richter Scale) occurred in a deep ocean trench off the west coast of Sumatra. The resultant tsunamis swept across the Indian Ocean, leaving a final death toll of 229,866. Earthquakes are natural events. We cannot prevent them. Yet as with so many natural events, it is only by looking at how people manage their environment that we can understand why they turn into disasters. On the underdeveloped coasts of Sumatra, where the mangrove swamps have been left intact, or on Thailand's Surin Islands, where coral reefs remain extensive and healthy, the worst force of the waves was absorbed and the death toll was minimal. On the majority of coastal areas such natural protection has been stripped away. Development and settlement have been allowed right down to the beachfront. It was in such places that the tsunamis inflicted the greatest damage and the greatest casualties.

In late August 2005 Hurricane Katrina began its journey across the Caribbean. The quality of storm defences, as well as human settlement patterns, determined just how much damage Katrina wrought. The interaction of human and physical forces was, perhaps, most clearly seen in New Orleans. The settlement of an area that is below sea level and protected by inadequate flood barriers, combined with the existence of cultural and economic forces that concentrated poor African-Americans within the most vulnerable areas, created images of social breakdown within the world's superpower. The rising levels and warming of the oceans provide conditions for an increased incidence of such high intensity storms. Global climate change offers us another probable link between human activity and natural disaster. As Halford Mackinder argued in 1887: 'Man alters his environment, and the action of that environment on his posterity is changed in consequence'.[68]

Modern societies look for technological solutions to their problems. Tsunamis do not cause us to stop building in vulnerable areas but to build better early warning systems. Flooding in low-lying areas does not mean people stop living in such places but provokes them to make their flood defences bigger and higher. Similarly, knowing that the burning of fossil fuels is damaging the environment is not leading nations to significantly reduce energy consumption but, rather, to find other energy technologies, such as cleaner engines and nuclear power. The 'technical fix' is the modern way. Yet as Stephen Schneider and Penelope Boston suggest,

> *whether to risk radically modifying the Earth in pursuit of human goals is not a scientific question per se; rather it is a fundamental political value choice that weighs the immediate benefits of population or economic growth versus the potential environmental or societal risk of a rapidly altered Earth.*[69]

As long as we imagine ourselves outside of, or above, nature, the risk involved in 'radically modifying the Earth in pursuit of human goals' will probably seem well worth it. If we are beyond nature then we can 'fix' our way out of anything. Yet today we know far more clearly than before that this is a delusion; that we are part of nature. Although it can sometimes *appear* as if we live in an entirely human-made world, we have come to understand that this is a fragile conceit. We are recently evolved animals living on a cooling remnant of a small star. If we make the planet uninhabitable for ourselves the world will not stop turning.

Geographical Obsessions: Urbanisation and Mobility

Introduction

The twin pillars of modern geography are environmental and international knowledge. However, geographers have a number of more specific obsessions. None are more important than the two processes I shall be looking at in this chapter, *urbanisation* and *mobility*. Both have ancient roots. Both also reflect two of the most powerful social dynamics of the modern age, the rise of cities and the growth of large-scale international migration.

Urbanisation and mobility are projects and problems that lie at the heart of what used to be called the 'rise of civilisations' and is now, increasingly, depicted as the formation of global modernites. In this chapter I shall be charting the rise of the 'modern urban view point'; a perspective that sees urbanisation and migration between sedentary communities as fundamental to human progress.

Yet it is a strange kind of progress. For urban societies are haunted by nostalgia and suspicious of their own creation. 'Why are men huddled together in unmanageable crowds in the sweltering hells we call big towns?' asked the English socialist, William Morris, in 1884.[1] Is Morris's rhetorical query a radical or a conservative one? Maybe both. What is clear is that of all the *places* to attract the attention of geographers the city is likely to remain the most important. The story of the city, its forms, structures and meanings, is constantly retold. And each retelling faces the paradoxical nature of the urban achievement. The city is cast as the most remarkable example of humanity's will to control the landscape and impose order. But it also offers us a landscape marked by social segregation and that most modern of experiences, alienation.

Presenting the City

We live in an urban world. Half the world's people live in towns or cities.[2] Moreover, the urban population is growing far faster than the rural population. By 2030 the proportion of the world living in towns and cities is predicted to be 60 per cent.[3] Urbanisation rates are highly uneven. In 2005 in the least urban country, East Timor, only 8 per cent of people lived in towns or cities.[4] In Sub-Saharan Africa the figure was 38 per cent, in Asia (not including the Middle East), 37 per cent. In contrast, in the same year 82 per cent of South Americans were urbanites.[5] An ambition of the Chinese government is to ensure that the country is predominately urban by 2020. As is so often the case in the history of urbanisation, the 800 million urban residents required to make this happen (up from 502 million in 2002), are understood to be integral to the achievement of a wealthy, forward looking, society.[6] One Chinese government official, Wang Mengkui, explained that

> *A country where most of the population is in poor or remote villages will not be a modern and developed nation. Our urbanisation rate [of 39 per cent] now is equivalent only to that of the UK in the 1850s, that of the US in 1911 and that of Japan in 1950.*[7]

The city can appear to be a recent creation. Yet beneath the glass towers of the twenty-first century metropolis lie the remains of far older settlements. Urbanism is a modern ideology. But it has a long heritage.

The first city is conventionally identified as Ur. Some 5000 years ago a cluster of villages in Southern Mesopotamia (in modern Iraq) came together to form a larger settlement. What is to be gained from living in this way? Ur provides the first answer. Just before its foundation a change of climate had placed new demands on irrigation: a bigger and better irrigation system was needed. By grouping and dividing their labour the people of Ur were able to build and maintain a new agricultural infrastructure that benefited them all. The same principle has been at work ever since. The city allows us to manage and employ more resources, both practical and cultural, than other kinds of places.

The city helps people do things. Yet there is an almost equally old understanding that people must live in a certain way in the city: the city is controlled by people but also controls people. This is nicely illustrated when we look at the ancient Chinese ideograph for city,

The ideograph depicts a man kneeling beneath an enclosure. The enclosure is the wall of the city. It also represents the symbolic power of the city as a ceremonial and political centre. The man is kneeling beneath these powers: he is submitting to the city. The urban order demands that landscapes be tightly disciplined, moulded to a system that allows human control and demands human subservience.

The city has long been associated with civilisation. Strabo's depiction of the world's various nations returned repeatedly to two themes, their physical environment and their cities. The rise and splendour of cities were taken by Strabo to indicate a flourishing civilisation. By contrast, the absence of cities connoted barbarism while deserted cities evidenced 'madness' and decay. The process of urbanising and, hence, civilising and colonising rural, 'undeveloped', societies is one that can be traced from imperial China and Rome through to the globalisation of Western city forms seen over the past one hundred years. The geometric, gridiron style of planning that is characteristic of modern cities allows for the quick and easy imposition of a new colonial order. As Lewis Mumford explains, the city grid reached its ancient apogee within the Roman empire:

> *The standard gridiron plan ... was an essential part of the kit of tools a colonist brought with him for immediate use. The colonist had little time to get the lay of the land or explore the resources of a site: by simplifying his spatial order, he provided for a swift and roughly equal distribution of building lots.*[8]

The Roman surveyor planning a new settlement would position himself on the highest point overlooking the site. Once there he would use a *groma* – a stick with cross arms holding two plum lines – for planning straight lines and surveying squares and rectangles and determining the centre of the colony where the forum would be sited. Town planning was not invented by the Romans, nor was the gridiron plan, but they extended and refined it. Indeed, the grid of the central part of the city was overlaid in some Roman towns with another that stretched into the countryside. Called 'centuriation', it divided land for settlers all around the city, mapping the city's spatial form across a wider urban region.

These early attempts to impose order on the landscape contain an ideal of control but also of proportion, beauty and harmony. The latter aspiration reminds us that the city has also been a repository of humanity's utopian ambitions. A kind of perfection is found in plans for new cities, which are as regular as a gaming board or as nature's own symmetries. Palma Nuova, founded in the state of Venice in 1593, appears to combine both types of geometry (Figure 3.1). Its regular streets, efficient fortifications and defensive shape bring together military and aesthetic ambitions.

Town planning is both prescriptive and descriptive. An ideal urban structure is presented to us at the same time as an analysis

Figure 3.1   Palma Nuova, founded 1593

Source: Johnson, 1967

of how people make use of urban space. As this implies, town planning struggles against the past; against organic and haphazard amalgamations of form, network and place built up over the generations. But it is also pitted against the chaotic present; more specifically frenzied, disorderly, expansion. Indeed, contemporary town planning is often seen as a reaction against the unregulated expansion of the nineteenth century city. In *Urbanisme* (1925) the Swiss architect Le Corbusier warned that 'the Great City, which should be a phenomenon of power and energy, is today a menacing disaster, since it is no longer governed by the principles of geometry'.[9] The remedy has come in the shape of state bureaucracies that are supposed to oversee local and national planning. Aligned with new building technologies, these institutions both encourage and combat an expectation that the city can be rapidly reshaped according to changing cultural attitudes, political will and economic demands. The need to control and order the city exists alongside the need to allow continuous urban transformation.

The life expectancy of most new buildings is very short. It has been estimated that by 2030 over half of all the buildings in the USA will have been built after 2000.[10] This openness to change can overwhelm other considerations. Indeed, it can sometimes seem as if planning has become an allusive dream; that the city is now defined by a constantly renewed sense of impermanence, noise and disharmony. The perpetual flux of the modern city has its own fascination. Trying to understand how the city works, its divisions and its injustices has become a recognised specialism (urban studies) and sub-specialism. Urban geographers are joined by urban sociologists, urban planners, urban political scientists, all vying to capture the fluid life of the city. Their models have traditionally sought to rationalise a confusing reality. Indeed, they draw on the kind of geometries that have symbolised order and harmony for millennia. The best known of these models was devised in the swiftly changing landscape of early twentieth century Chicago. The most important feature of Ernest Burgess's 'Concentric Theory of Urban Structure' (Figure 3.2) is the 'zone of transition'.[11] This area surrounding the Central Business District receives new immigrants and is characterised by 'unstable' social relations. The further one moves out, the more prosperous the residents (and the longer the journey into work). One can see from the second of the two diagrams that while new immigrant groups are confined to the centre, as they become more assimilated they too move out. In the first instance they migrate to outlying enclaves, until eventually they are fully absorbed and disappear as a discrete spatial entity.

Burgess was a pioneer and his model has been criticised and revised many times. His model reflects a vigorously growing US city. It is based on a sanguine view of the longevity of racial division and a presumption that all social groups share the same attitudes

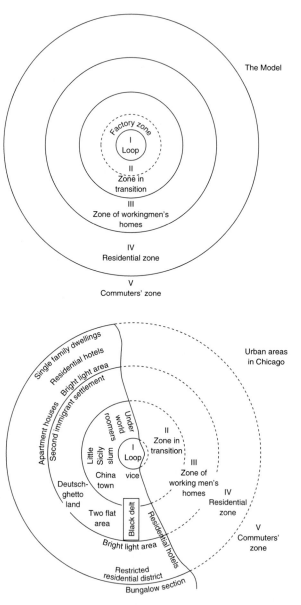

Figure 3.2   The Concentric Theory of Urban Structure

Source: Burgess, 1925

to urban living. Yet despite its questionable assumptions, Burgess offered a vision that captured realities that were still being noticed by later generations of urban theorists. Thus, for example, in a study of housing choices in Birmingham, England, Jon Rex found that

> *The persistent outward movement which takes place justifies us in saying and positing as central to our model that suburban housing is a scarce and desired resource.*[12]

Urbanisation has also moved the city into the centre-ground of political and economic debate. During the period when modernity was taking two political roads, communist or capitalist, the modern city became either the capitalist city (of the West) or the communist city (of the East). However, the latter was not an object that held the attention of the Marxist scholars who, in the 1970s and 1980s, came to dominate urban studies in the West. Despite their narrow frame of reference, this group staged Marxism's most impressive recent intellectual reinvention. Within academic circles, understanding the city as an arena of capital/ism became a leitmotif of its rigorous analysis. The best known Marxist geographer is David Harvey. Harvey portrays the built environment as a 'secondary circuit' for capital; an arena where money may be employed and deployed when profit rates are declining elsewhere in the economy. Thus he presents urban development and regeneration as attempts to avoid economic crisis:

> *Each of the global crises of capitalism was in fact preceded by the massive movement of capital into long-term investment in the built environment as a kind of last-ditch hope for finding productive uses for rapidly over-accumulating capital. The extraordinary property boom in many advanced capitalist countries from 1969–73, the collapse of which at the end of 1973 triggered (but did not cause) the onset of the current crisis, is a splendid example.*[13]

In Harvey's model the city's expansions and crises are rooted in the management of the accumulation of capital. The expansion of consumption is also tied into this process. Thus,

> *A variety of strategies emerged for stimulating consumption, not least*
> *of which was a set of fiscal and monetary policies designed to acceler-*
> *ate and enhance the suburbanization process.*[14]

There is a compelling boldness and analytic clarity to Harvey's
explanation. However, over the last three decades, Marxist
approaches to the city have been widely criticised both for reducing
everything to economics and, relatedly, for failing to appreciate the
city's patterns of ethnic, gender and sexual diversity. Harvey has
met these concerns head on, interpreting them as symptomatic of
the ability of a new 'post-modern capitalism' to assert a fragmentary
anti-politics of personal recognition and identity.[15] As this counter-
attack implies, one of the things that Harvey has found galling
about his critics is that they appear to resist the idea that the city
can or should be subject to abstract modelling. Harvey sees a capitalist order in the modern city. But he is also defending an age-old
geographical imperative: the desire to *impose and find order* in the
world.

We have come a long way from Ur. In recent years the remains of
Ur, the first city, have been damaged by US forces occupying Iraq.
But our thirst for the urban appears unquenchable. The city continues to evidence and inspire a yearning for order and control. It represents one of humanity's most notable achievements and its most
magnificent imposition upon nature. Yet, as we shall now see, it
remains on object of suspicion.

Urban Critics

It is a paradox of civilisation: cities are expressions of order and
rationality yet this ability to control and contain is widely identified as a fall from grace. Rural life is associated with marginality.
But urban life is found to be artificial and oppressive. In Europe
these ideas have long been mapped onto a romantic willingness to
criticise European decadence by comparing it to the pristine morals
of societies without industry or towns. Referring to an incident
from 1562, the sixteenth century French essayist Montaigne

recalled that three 'natives' of the Americas 'leaving the gentleness of their regions to come and see ours' were shown round Rouen: 'The King had a long interview with them: they were shown our manners, our ceremonial and the layout of our city'.[16] Yet Montaigne tells us that these 'men of the Other World' are most struck by the inequality of 'men bloated with all sorts of comforts' living side by side with those

> *emaciated with poverty and hunger: they found it odd that those des-*
> *titute halves should put up with such injustice and did not take the*
> *others by the throat or set fire to their houses.*[17]

The stark proximity of economic divides in the city has traditionally ranked as one of its most keenly noted problems. The prospect of living amid a crowd of strangers has also long troubled social critics. The alienated interpersonal relations associated with the city began to be systematically analysed by German sociologists in the early twentieth century. For Georg Simmel and Ferdinand Tonnies the growth of urban life meant a move from organic and informal association to legally mediated and instrumental forms of connection between strangers.[18]

> *The reciprocal reserve and indifference and the intellectual life condi-*
> *tions of large circles are never felt more strongly by the individual in*
> *their impact upon his independence than in the thickest crowd of the*
> *big city. This is because the bodily proximity and narrowness of space*
> *makes the mental distance only the more visible. It is obviously only*
> *the obverse of this freedom, if under certain circumstances, one*
> *nowhere feels as lonely and lost as in the metropolitan crowd. For here*
> *as elsewhere it is by no means necessary that the freedom of man be*
> *reflected in his emotional life as comfort.*[19]

The ambivalence of the city's freedoms, which Simmel addresses in this passage, has also been a central theme of one of the most provocative contemporary urban critics, Richard Sennett. Given the theological function of many early cities, it is interesting that Sennett has arrived at a theory of metropolitan redemption, in which urban alienation is cast as a necessary escape from Paradise; an exile from organic community that enables people to respect others as different from themselves. It is a strange but compelling vision, for it injects a mystical content into the mechanical heart of the modern city:

> *The city brings together people who are different, it intensifies the complexity of social life, it presents people to each other as strangers. All these aspects of urban experience – difference, complexity, strangeness – afford resistance to domination. This craggy and difficult urban geography makes a particular moral promise. It can serve as a home for those who have accepted themselves as exiles from the Garden.*[20]

Perhaps, though, there is something a little too accommodating about Sennett's acceptance of the impossibility of human solidarity. The angriest critics of modern urbanism have never given up on this perennial hope. None were angrier than the Parisian revolutionary geographers who, from the late 1950s, called themselves situationists. A few years before his suicide, this group's leading figure, Guy Debord, bitterly complained that: 'Whoever sees the banks of the Seine sees our grief: nothing is found there now save the bustling columns of an anthill of motorized slaves'.[21] Debord articulated a hatred of all attempts to accommodate people to alienation. And he saw in the clean, homogeneous domestic boxes of the post-1945 building boom attempts to do this precisely that. For Debord, new urban developments signalled the eradication of popular memory and hence, insurrection. 'The "new towns" of the technological pseudo-peasantry', he spat, have as their motto: '"On this spot nothing will ever happen, and *nothing ever has*"'.[22]

Western worries about the city represent only a partial window onto what has become a global set of concerns. By 2005 the proportion of the world's urbanites living in developing countries had grown to 43 per cent.[23] Non-Western critics of urbanisation bring new dimensions to the debate, such as the way the city can become a locus of Westernisation and integration into the global economy. These forces may of course intersect, a possibility that has provoked the urban geographers Stanley Brunn and Jack Williams to suggest that:

> *change (in less developed countries) is likely to take the form of Westernisation, since it is the West, the more developed countries, that developed the lead in creating the modern industrial city and the*

life-style that goes with it. To be sure, there are ample signs of this Westernization (some might call it 'homogenisation' or 'international-isation') of the world's major cities, in the form of sky scrapers and modern architective, the automobile society, Western high technology, advertising, etc.[24]

In many parts of the world, though most especially in the Americas, Western colonisation was closely tied to the establishment of new cities. Indeed, Burkholder and Johnson argue that 'conquistadors and early settlers defined their colonisation of the New World by founding cities, which they saw as their link with European civilisation and culture'.[25]

The association of the urban with colonial authority meant that anti-colonial movements were sometimes articulated through an anti-urban, 'pro-peasant' politics. Anti-colonial nationalists in Asia and Africa often looked to the 'uncorrupted' countryside as the site of an uncolonised racial and national essence.[26] The desire to seek out political authenticity in the countryside occasionally led to a sense that the urban was, in itself, an alien form. The most notorious expression of this line of argument was offered by the militantly anti-urban Khmer Rouge regime of Cambodia (1975–1979). The regime's leader, Pol Pot, instructed: 'Don't use money: don't let the people live in cities'.[27] With the cities emptied and money forbidden it was hoped that a truly communist, purely peasant, Khmer (the Khmer are the largest ethnic group in Cambodia) society could begin. It is an extreme example; the most violent end of a spectrum of anti-urban suspicion.

Anti-urbanism reaches from the sadistic hostility of the Khmer Rouge to the humane pastoralism of Ghandi and the Indian poet Rabindranath Tagore. For Tagore the West is a mechanical, officious civilisation, the antithesis of the organic culture found within Asia. This distinction mapped onto another: the West was essentially urban and spread itself around the world by way of urbanisation. Authentic Asia, by contrast, was to be found, not in any particular nation, but in the countryside. 'The modern towns', Tagore wrote in *Creative Unity* in 1922, 'are everywhere the same, from San Francisco to London, from London to Tokyo. They show no faces, but merely masks'.[28] It was a theme that Tagore returned to on several occasions, for he saw the Western relationship between town and country spreading throughout Asia: 'unlike a living heart, these cities imprison and kill the blood and create poison centres filled with the accumulation of death'.[29]

It might be expected that Tagore's ghastly portrait of the city would have faded away as urbanisation became so dominant as to be unassailable. However, suspicion of the city is constantly finding new champions. The anti-Western rhetoric of Islamic radicalism at the end of the twentieth century often invoked the corrupt nature of the Westernised city (and indeed, cities as such).[30] In less apocalyptic and confrontational terms, critics of the deplorable living conditions of urban workers in the booming cities of South and East Asia wonder too at the long-term benefits of megalopolis.[31] As we enter a period in which urban life will be the norm across the globe, the debate about the nature of urban life is likely to grow even more intense.

Decentering the City

The city is dramatic. It is the central stage of modern history. And it is talked about in spectacular terms: cities are announced to be 'dead', to be 'reborn', to be 'exploding' and being 'reinvented'. The ability of the city to gobble up surrounding settlements into 'conurbations' (a word invented by Patrick Geddes in 1915) was a particular concern of one of its most wide-ranging chroniclers, Lewis Mumford. Writing in the late 1950s, Mumford noted that,

> *If no human purposes supervene to halt the blotting out of the countryside and to establish limits of the growth and colonization of cities, the whole coastal strip from Maine to Florida might coalesce into an almost undifferentiated conurbation. But to call this mass a 'regional city' or to hold that it represents the new scale of settlement to which modern man must adapt his institutions and his personal needs is to mask the realities of the human situation and allow seemingly automatic forces to become a substitute for human purposes.[32]*

In 1961 Jean Gottmann coined the word 'Megalopolis' for this urban strip.[33] Within the modern, mobile, economy, 'real places', with their associations of organic community and stubborn local attachment, become awkward points of immobility. Arif Dirlik argues that 'places' have been rendered 'into inconveniences in the path of progress to be dispensed with, either by erasure or, better still, by rendering them into commodities'.[34] Dirlik's reference to commodification reminds us that the 'death of place' allows ex-places to be turned into brands, available for marketing and instant reinvention (a process that is often accompanied by the opening of a 'heritage centre').

The titles of Edward Relph's *Place and Placelessness,* Marc Auge's *Non-Places* and James Kunstler's *The Geography of Nowhere,* indicate a similar set of worries.[35] For Auge, non-places are 'where people coexist or cohabit without living together'.[36] It appears that a dissatisfaction with 'placelessness' has helped created counter-currents, such as self-consciously distinct 'urban villages' and a flight to the countryside. Ironically, the decision of increasing numbers of the affluent to live in 'real places' may have turned into yet another force threatening community life, by turning villages and desirable urban enclaves into the preserve of a moneyed elite.[37]

However, the decentring of the city has also been presented in less gloomy terms. The new, centreless metropolis is associated by postmodern writers with a liberation from rigidity and tradition. Indeed, the possibility of any kind of modelling or analysis of the decentred city has been called into doubt. Loretta Lees reports that, 'Unlike the Chicago School's concentric zone model' the 'new postmodern city is almost impossible to represent'.[38] In even bolder terms James Heartfield informs us that 'London is over'.[39]

Since models of the city have always been more to do with the human will to find and impose order, giving up on representing the city because one believes, as Lees does, that 'cities are more complex than ever before', is unlikely to be a long-lasting intellectual trend. Swooning before complexity is a reflection of a decay of analytical purpose. Marx described the socio-economic basis of this state of agitation:

> *Constant revolutionising of production, uninterrupted disturbance of all social conditions, everlasting uncertainty and agitation distinguish the bourgeois epoch from all earlier ones.*[40]

Another reason Marx is pertinent is that just over 30 per cent of the world's population live in countries that have tried and now abandoned the Marxist alternative to 'bourgeois' revolution. In this context, the postmodern penchant for everlasting uncertainty appears akin to the ideological crowning of late capitalism. However, a final, paradoxical, reason that the name of Marx is relevant concerns the fact that, in academic geography, a neo-Marxist hostility to 'place-bound nostalgias'[41] has framed much of the recent debate about place. Thus, terms such as 'defensive' and 'reactionary' have been employed to account for places thought to be inadequately fluid and changeable. The most eloquent of such critics is Doreen Massey. Her 1993 essay 'A global sense of place' has become a touchstone for those who wish to challenge claims that particular places have rooted, or in some way organic, identities and histories.[42] In this piece Massey celebrates Kilburn, in London, as a politically progressive place because it permits 'multiple identities'. Massey describes her attraction to Kilburn through a left-liberal semiotics of everyday space. Her 'walk down Kilburn High Road' reveals 'many an empty space of a wall ... adorned with the letters IRA [Irish Republican Army]'; her chat with the newsagent finds a 'Muslim unutterably depressed by events in the Gulf, silently chaffing at having to sell the *Sun*'.[43]

Massey's walk leaves me disoriented. Although it is supposed to be affirming diversity, her moral *sorting* of good and bad messages from the landscape is hard to miss. It appears that any possible tension between Massey's affirmation of 'multiple identities' and her strict left-liberal sympathies is resolved by the simple act of not seeing or listening to 'reactionary' messages.

A more plausible challenge to 'nostalgia' for organic, rooted senses of place emerges from arguments about the way new communications technologies are undermining the importance of proximity. In *City of Bits,* William Mitchell suggests that,

> communities increasingly find their common ground in cyberspace rather than terra firma ... Today as telepresence augments and sometimes substitutes for physical presence, and as more and more business and social interactions shift into cyberspace, we are finding that accessibility depends even less on propinquity, and community has become unglued from geography.[44]

This 'end of geography' argument is reinforced by Gottdiener and Budd: the notion that 'we need proximity and density for cities and suburbs to work', they write, is 'quite out of date in the information society'.[45] In comes the Internet, out go the neighbours. Postmodern urban critics seem to relish the decline of place-based solidarities. Yet, it is worth recalling that classical theorists of the urban condition, such as Simmel and Tonnies, questioned the conflation of 'proximity' and 'community' many years ago. In Louis Wirth's term, 'The contacts of the city may indeed be face to face, but they are nevertheless impersonal, superficial, transitory and segmental'.[46] The *possibility* of community in the city has been an axis of debate in urban studies for over 100 years. As this implies, the advent of 'telepresence' can be represented, not as a departure from, but as an intensification of, processes already firmly identified with modern urban life: namely, social fragmentation, alienation and superficiality.

One final thought. For I cannot help noticing that it is one of the most geographically footloose of professions, namely academia, that appears to produce the most determined critics of 'place-bound nostalgias'. If one is in a hyper-mobile, high status, occupation, 'place bound' community may, indeed, appear to be an irrelevance, or even a conservative anachronism. The view from less mobile and less affluent parts of society is likely to be different. Given that the last century witnessed an unprecedented level of destruction of established communities – through ethnic cleansing, military attack and large scale redevelopment programmes – the academic 'critique of place' strikes me as oddly disconnected. Perhaps academics are not well *placed* to produce engaged geographical knowledge. The question of who gets to produce geography is one I shall be returning to in Chapter 5.

Mobilities

Mobility is one of the keywords of contemporary life. It is imagined to have an intrinsic value. To be moving is be living, energetic and 'going places': to be static is to be moribund and passive. Given the environmental crises associated with mobility noted in the previous chapter, these associations may change. However, for the time being, flying up and away seems akin to personal liberation, and being footloose and flexible is an accepted corporate ideal. The three themes that dominate discussion of mobility are migration, mobile economies and travel.

Migrations

Ours is an age of migration. Of a certain type. Few truly nomadic communities still exist. Modern migration does not reflect age-old, cyclical patterns. It typically involves specific, often personal, circumstances that provoke a movement from one settled place to another. By far the largest of all migratory movements over the last few hundred years has been the drift from the countryside to the city. However, migration has also become associated with flows between nations. Looking at international migratory trends over the past 300 years one of the things we notice is the shift away from transcontinental migration to undeveloped countries and towards migration to closer, richer nations. Observing this phenomenon, Goran Rystad explains that

> Today's intraregional migration differs from yesterday's transcontinental migration to such an extent that whilst the migration waves of the eighteenth and nineteenth centuries wandered from richer to poorer countries, at least in terms of the level of economic development, current migration waves travel from less developed countries and regions to those more highly developed.[47]

The scale and nature of migration to the Americas from the sixteenth century onwards, both forcible (in the case of African slaves) and voluntary (in the case of European settlers), are historically unique. Today the indigenous population in Brazil and the USA (the most populous American nations) is under 1 per cent and the most important form of migration to affect the Americas is internal, notably the movement of Latin Americans to the USA. The Hispanic population of the USA has grown to 14 per cent (in 2004).[48] It has also been Hispanics who have, more than other groups, been responsible for the rise in the size of the foreign-born labour force, from 5.3 per cent in 1970 to 14.7 per cent in 2005.[49]

Why do people migrate? For many it is not a matter of choice. Slavery and deportations have been responsible for vast flows of people. The European slave trade transported about 10 million people from Africa.[50] More recently, communist regimes built up an indelible record of forced migration. Stalin transferred numerous ethnic and social groups, both because they were deemed to be politically suspect and in order to colonise sparsely settled areas of the USSR. Two of the largest transfers were of half a million ethnic Germans (1941–2) and the entire Chechen and Ingush nations (in all, 484,000 people) in 1944.[51]

However, the majority of migration contains an element of choice. One of the first people to analyse why people move was Ernest Ravenstein. In 1889 he argued that,

> *Bad or oppressive laws, heavy taxation, an unattractive climate, uncongenial social surroundings, and even compulsion (slave trade, transportation), all have produced and are still producing currents of migration, but none of these currents can compare in volume with that which arise from the desire inherent in most men to 'better' themselves in material respects.*[52]

Yet the decision to migrate is usually a difficult calculation. Even when the promise of work seems secure many other worries arise, such as how long employment will last and whether income can be found for other members of one's family. Another problem migrants encounter is changes in attitude in host countries. Policies towards migration seem to be especially politically volatile. The to'ing and fro'ing between wanting and not wanting migrants in western Europe may be summarised as follows:

Open Gate: 1860–1914 (with free immigration and
extensive emigration)
Shut Gate: 1914–1945 (immigration restriction and
alien control)
Open Gate: 1945–1974 (relatively free immigration and
direct recruitment of a foreign workforce)

Shut Gate/Open Gate: 1974 – (immigration restriction for non-EU citizens; increasing mobility of EU citizens within an enlarging EU) (adapted from Hammer, 1990[53])

In Europe, as in the Americas, we see a broad tendency towards international migration becoming contained within a particular region. The significant flow of citizens of former British, French and Dutch colonies into Europe seen in the 1950s and 1960s has been curtailed. Migration policies are increasingly tailored to encourage only the most economically active from arriving. Indeed, in Europe, and to an even greater extent in the USA, whole sectors of the economy have come to rely on low-wage, non-unionised migrant labour.

Immigration has become associated with 'flexible' economies. This flexibility also has social dimensions, particularly for migrants themselves. Immigrants are often supporting dependants back home. This pattern produces chains of remittance, dependence and care stretched over large distances. Moreover, although migrants are still predominately male, women are increasingly being pulled into the circuit of migrant labour, in part, because of changes in the division of labour between men and women in the West. One consequence has been that increasing numbers of affluent North Americans and Europeans buy-in their child care from foreign women, who in turn have to make complex arrangements for the care of their own relatives back home. Reviewing case-study evidence of this relationship from San Diego, Adrian Bailey finds that:

> *While middle-class San Diegans sourced domestic work locally, their employees were financially unable to meet their own care needs through subcontracting arrangement and instead wove complex international networks of caring that often meant prolonged periods of separation between them and their children/extended family, and increased backwards and forwards mobility across the border.*[54]

Mobile Economies

Trade has always been of interest to geographers. Ancient geographers routinely itemised the principle products and exports of the countries and cities they depicted. Trade connects places. In many cases, it creates places. The modern era has seen a growth in the mobility, speed and global nature of economic life. Indeed, in the late twentieth century this momentum increasingly provided the central test for national economies. The question was 'Can we keep up?' and it was asked of communism in the ex-Soviet Union and China, of 'developing economies' struggling to diversify, and of Western economies not wanting to be outpaced by societies with lower labour costs. Despite late attempts to open up the economy, the highly centralised economic model developed in the Soviet Union proved too slow and unwieldy to 'keep up'. Chinese communism survived as a political institution but only by abandoning communist economics. Today, at the start of the twentieth century, we have a highly integrated, global market economy. Thanks to the Internet and trade liberalisation, anyone can pull out a mobile phone or laptop almost anywhere in the world and order and sell goods and services and conduct negotiations with almost anyone else anywhere in the world.

In trying to understand how this new hyperactive global economy came into being we can isolate a number of enabling factors. I have already mentioned one, the death of communism. Not only did this bring vast new populations into the market system it removed the possibility of a non-capitalist economic option. The fact that there is only 'one game in town' makes a difference: today, without the choice of an alternative, societies and individuals plunge fully into capitalism without hesitation or circumspection.

Alongside the collapse of a rival system, we can isolate three other drivers towards economic mobility: the rise of international regulatory institutions; transnational corporations; and technological change. I shall briefly introduce each of these in turn.

It is one of the paradoxes of the mobile global free market economy that it relies on ever stricter international and national governance to keep it moving. It is a paradox rooted in capitalism's own inherent tendency towards short-term decision-making and the formation of monopolies. The 'free market' exists thanks to the protection of transnational and national public bodies. The second half of the twentieth century saw, for the first time, the foundation of truly transnational economic global regulation. The Agreements signed at Bretton Woods in 1945 were designed to stabilise war-torn economies. In founding both the World Bank and International Monetary Fund (IMF) and then, in 1947, the General Agreement on Tariffs and Trade (GATT), Bretton Woods established a regime of governance for free trade. While the World Bank and IMF provided funding and advice, the GATT created a template of practice. The GATT was based on four principle rules: (1) that there should be a 'level playing-field' and, hence, an end to preferential trading terms; (2) that there should be reciprocity of tariff reductions; (3) that there should be transparency of trade measures; (4) that there should be an end to protectionism and dumping practices. These principles were not legally binding. When GATT was replaced by the World Trade Organisation (WTO), in 1995 (which has 149 members, as of December 2005), it was given legal teeth: non-compliance now invoked investigation, sanctions and fines. The most significant recent member of the WTO, in terms of both political significance and volume of trade, is China, which joined in 2001.

This global institutional structure is supplemented by regional free-trade alliances. Some 40 per cent of world trade takes place within regional trade agreements.[55] In terms of value of trade, the North American Free Trade Agreement (NAFTA) (founded 1994), the Association of Southeast Asian Nations (founded 1997) and the European Union (EU) are the most important. The trajectory followed from its foundation, in 1957, as the European Economic Community, to the closer political and social integration marked by the inauguration of the European Union (in 1992), has established a wider expectation that regional economic integration is a first step along a longer path. However, at the heart of the EU lies trade. The introduction of a common currency, in 2002, represented the EU's greatest single achievement. Coupled with considerable steps towards the unrestricted mobility of labour and capital, the EU is well on its way to becoming a single economic unit.

The formation of regional free-trade alliances has, in part, been promoted by the fear of some national governments that they are too small, and lack the economic clout, to negotiate with transnational corporations (TNCs) on equal terms. The sales of the largest TNCs are far bigger than the Gross Domestic Product of most nations. In a combined list, from 2003, we find that Wal-Mart at US $ 256.33 billion sales is the largest TNC, and comes above Austria's GDP of US $ 251.46 billion (further down the list we find Turkey at US $ 237.97 billion, followed by BP's US $ 232.57 billion and ExxonMobil at US $ 222.88 billion).[56] Moreover, the global reach of TNCs allows them a footloose capacity to pick and choose where, when and how to design, produce, market and acquire. This ability is, in part, predicated on the disappearance of what is sometimes called the 'old international division of labour'. In this essentially colonial model, the 'periphery' supplied raw materials, which were turned into finished products in the industrial 'core' (often to be sold back to the periphery). From the 1960s onwards the outflow of investment from the West and the rise of manufacturing capacity in East Asia signalled that a new arrangement was emerging. In the 1970s Folker Fröbel et al. argued that what they called a 'new international division of labour' was developing.[57] For Fröbel et al. the population resources freed by agricultural intensification (i.e, the Green Revolution), and the subdivisions of labour possible in new factories (which allow little or no worker training), have combined with innovations in transport technology (notably containerisation) to reorientate the world economy. Fröbel et al. suggested that the periphery should now be associated with mass manufacturing and the core with management, research and high level skills and manufactures.

With hindsight, Fröbel's et al's 'alternative' model appears too hesitant. For what was changing was not simply the capacity of 'the core' and 'the periphery' but the division of the world in binary terms. The *flexible capitalism* of the last two decades makes use of the Internet and computer technology to create new opportunities for footloose, 'just in time', consumer driven corporations. The hiring and firing, movement and retraining of workers, combined with a growing tendency to contract out activities, have established a new de-centred business model. Robert Gwynne et al. in a recent study of new economic networks, explain that producers of such items as toys, clothes and footwear are 'essentially ... manufacturers without factories, at the centre of a highly flexible and global network of production distribution and marketing'.[58] This scattering of economic functions favours those environments, notably in East and South Asia, with a wide skill base and where worker protection is weak. The ultra-mobility of this new economic order contrasts sharply with the ability of a handful of Western capitals (notably, Tokyo, London and New York) to monopolise global financial services, and offer themselves as the 'pinnacle of the global urban hierarchy'.[59] However, in the medium to long term this lofty position is far from secure. As workers in the rest of the Western economy have learnt, in the modern world the centre does not hold.

The Age of Travel

A car for every purse and purpose.
(Alfred Sloan, President of General Motors, 1924)[60]

The notion that modern life is characterised by speed and mobility is inseparable from access to modern transport. Cars and aeroplanes are objects of fascination for geography. They change our relationship with both our immediate and our global environment. The environmental knowledge that geography offers also makes cars and aeroplanes objects of concern. They have come to symbolise a short-term and unsustainable disregard for both the atmosphere and the landscape. Indeed, it seems that our 'age of travel' is partly responsible for a weakening of people's attachment to their local environment. Describing the situation in Sweden, Olle Hagman writes,

> The 'local' community in which the places of living, working, recreation and leisure activities were located within walking or cycling distance has given way to a 'dispersed' society, shaped by and for private car ownership, in which the average travel length for adults in Sweden is approaching 50 kilometres a day.[61]

National travel surveys in Europe suggest that it is the Swiss who are most likely to be on the move at any one time. The average 88.8

minutes the Swiss spend travelling each day compare with a mere
13 minutes for Latvians (figures from 2000 and 2003 respectively).[62]
Detailed data for Scotland confirm just how rapid and significant
the increase in car usage has been over the past two decades. The
National Travel Survey for Scotland found that, between 2002–3,
the average adult Scot travelled 6,670 miles per year. Since 1975–6
this average has risen by 2,500 miles. Cars accounted for 86 per
cent of the increase, making up 4,000 miles of the total.[63]

Car usage is continuing to expand, particularly in the developing
world. The new horizon for car manufacturers is Asia. China is the
world's fastest growing automobile market. Since only three in 1,000
Chinese own a car (as of 2003) the potential is considerable.[64] In contemporary China car ownership is increasingly seen to be the ultimate symbol of both personal success and the benefits of a market
economy. Jinya Chen, the president of the Chinese branch of the
world's largest auto parts supplier, Delphi Corp, predicts that the
'boom will continue. The question is how big the boom will be ... In
China, most people think that having a car is a dream come true'.[65]

The past decade has seen air travel achieve even greater levels of
growth. International air travel has grown from something
restricted to a small elite to a prosaic expectation, at least among
many in the West. Demand for foreign travel is spreading rapidly.
It has been predicted that by 2020 the number of Chinese tourists
travelling abroad each year will reach 100 million.[66] As air travel
becomes a world phenomenon its environmental, socio-economic
and cultural consequences are likely to become matters of even
more intense and widespread concern.

Quite what impact the expansion of tourism will have on host
societies, and tourists' perceptions of themselves and the world, is
especially intriguing. Despite its mass popularity, tourism retains a
certain hauteur. The world becomes an arena of 'experiences' to be
consumed. And as speed and ease of travel increase so too do the
anxieties about the superficiality of this type of engagement with
the world. Dan MacCannell identifies the encounter between tourist
and destination as an *'empty meeting ground'*, devoid of any expectation of human contact or vulnerability.[67] The world has become
accessible, available. Yet our ability to encounter and respect people

and peoples who do not conform to the increasingly universal template of modernity may be diminishing. Modern travellers are prepared to go anywhere but they expect 'modern standards', even a home from home, when they get there. Perhaps the problem goes deeper still. The modern gaze is often blank, indifferent, flickering into life only before a cliché. In a prescient remark from 1922, Rabindranath Tagore worried that Western colonialism had become the paradigm for all human contact,

> *The modern age has brought the geography of the earth near to us, but made it difficult for us to come into touch with man. We go to strange lands and observe; we do not live there. We hardly meet men: but only specimens of knowledge. We are in haste to seek for general types and overlook individuals.*[68]

Conclusion

> *the crystallisation of mass about a nucleus is part of the elementary order of things.*
> *(Walter Christaller)*[69]

> *the history of mankind is the history of migration.*
> *(W. R. Bohning)*[70]

We are familiar with urbanisation and human mobility. So familiar that we may mistake them for inevitable, natural, processes. Both can, indeed, be traced back a very long way. Yet both are also expressions of modernity; more specifically, of the way people in the West have come to understand social progress. And each has emerged as an obsession for the modern geographical imagination in large part because each has come to represent the modern world, its desire to build bigger and move faster, its insistence on limitless human expansion. Today this is especially the case in the 'majority world', where urban growth and improved transport appear as the golden keys to development. People in the built-over, traffic-clogged, West are hardly in a position to object.

Both as global phenomena and as phenomena that enable us to see globally (by bringing a diverse world closer together), urbanisation and mobility offer a prospect of human alliance and mutual tolerance. This shimmering horizon is clouded, however, by some profound doubts. Urbanisation and hyper-mobility wrench us away from nature. They cause more and more of us to live disconnected from an intimate knowledge of place, landscape and environment. As we have seen, they have also provoked laments about their impact on human solidarity. As we build higher and travel further, a sense of loss clings to our ascent. Although many academic critics disparage the need for place-based communities, over the past 100 years maintaining community and a sense of place (or even just hanging onto the hope that such things are possible) has been a hard but necessary struggle for people uprooted and relocated. Continual deracination and the continual rebuilding and expansion of the urban realm are dramatic spectacles. They are the modern way: fascinating and mighty. But they are also troubling.

4 ●● Doing Geography

Introduction

Geography surrounds us. We drive, fly, and walk, over and through it. At the same time, it is distant: geography is the landscapes beyond the horizon, the intriguing distance between here and there. So far I have been answering the question 'What is geography?' by looking at what geographers are interested in. This chapter looks at *how* they undertake their work. There are some specific practices associated with geography. The best known is mapping. Other activities closely allied with geography are travelling, fieldwork and connecting the human with the natural landscape. There are many more. After all, if we accept that geography is the world discipline, then its methods will be countless. However, this definition also allows us to distil the four characteristic verbs of geographical practice – the four actions that allow the geographer to understand the world – to *explore;* to *connect;* to *map;* to *engage.*

Geographical study cannot be contained entirely by libraries and laboratories. It is about our world and demands that we get out into the world. This makes geography difficult to institutionalise. Geography wants to take children outside the school and into the streets and fields; it wants to take keyboard tappers out of their gloomy offices and into the rain or the sunshine. This yearning for physical sensation and real life makes geography an awkward customer in contemporary societies that are risk averse, remorselessly monitored and highly bureaucratic. But that might just be to its advantage.

To Explore

Geography is constantly urging us to step outside the door. Yet isn't exploration dead? '[W]e are now near the end of the roll of great discoveries', Halford Mackinder told the Royal Geographical Society in 1887, the 'tales of adventure grow fewer and fewer'. Thus it is, he added, that 'even fellows of Geographical Societies will despondently ask, "What is geography?"'.[1] Yet exploration was never simply about Victorian gentlemen venturing into 'undiscovered' lands. Exploration is a journey into the world in search of the new: it is a preparedness to physically encounter the world and study and learn from it. This definition allows us to identify three of the most important forms and issues in the practice of exploration today: fieldwork, travelling and knowing and the rise of the era of 'mass exploration'.

Fieldwork

To me, fieldwork is the heart of geography ... It renews and deepens our direct experience of the planet and its diversity of lands, life and cultures, immeasurably enriching the understanding of the world that is geography's core pursuit and responsibility.[2]

The concept of 'fieldwork' originated in land surveying. More broadly, fieldwork refers to investigations that take place in the same location as the phenomenon to be studied. One of the first uses of the term in this wider sense came in 1922, in Bronislaw Malinowski's study of Pacific island life, *Argonauts of the Western Pacific*.[3] However, despite this relatively recent etymology, the notion that the geographical study of a place should, at least in part, be based on work actually conducted there, is an ancient one. Indeed, the authenticity of geographers' accounts has traditionally been tested against their claims to have actually gone and seen for themselves.

Recently, the reliance on 'the field' as a test and verification of authenticity has worried some critics. The 'taken-for-grantedness' of the field strikes Heidi Nast as indicative of a muscular anti-intellectualism.[4] For Gillian Rose, fieldwork traditions 'construct access to knowledge of geography as a white bourgeois heterosexual masculine privilege'.[5] The 'bourgeois' tag is the least convincing of these four labels. With the aid of some British examples we can see that fieldwork came to be considered a necessary feature of a geographical education with the development of *mass* education. Some of the key figures in shaping school geography in Britain were also instrumental in asserting the core value of fieldwork. Thomas Huxley's influential early text *Physiography* (1877) was unequivocal in claiming that geography was a field-based science. There is a democratic, earthy quality to Huxley's insistence that geography is 'to be learned in the village and countryside, not read about in books'.[6] Huxley's approach influenced the expectations of school inspectors, who began to insist that the schoolground and the wider locality must be seen as resources for geography lessons. As Teresa Ploszajska explains,

> *fieldwork, whether in familiar local surroundings or distant unknown areas, was widely considered to be a means of encouraging children to recognise geography as this body of knowledge pre-eminently concerned with the real world.*[7]

Geography was becoming established as the one slot on the curriculum that allowed students to escape their stuffy classroom. It has guarded this ambition ever since. Three of the most influential British geographers of the last century left lasting legacies for fieldworkers:

S.W. Wooldridge, co-author of *The Spirit and Purpose of Geography* (1951) with W. Gordon East, helped establish, in 1943, the Council for the Promotion of Field Studies (subsequently the Field Studies Council), which today provides a network of 17 field study centres across Britain for school children.

Alfred Steers, who walked the coastlines of England and Wales in 1944, documenting coastal environments. Steers was influential in the establishment of the Nature Conservancy Council, which, in turn, led to the creation of Sites of Special Scientific Interest, designed and designated for both conservation and fieldwork. Today there are over 4,000 of these in England, covering 7 per cent of the land area.

Dudley Stamp, who began the Land Utilization Survey in 1930. In what remains the largest example of applied geography in Britain, a small army of school children was put to work to record the land use of every field in Britain. The results (eventually published as *The Land of Britain: Its Use and Misuse*, 1948) were used to plan Britain's food needs, both during the Second World War and in the immediate post-war period.

Fieldwork remains a core component of geographical practice. However, there is a growing connection between scientific and purely educational forms of fieldwork, and the development of leisure and voluntary based environmental and international encounters. 'Eco-tourism', 'geo-tourism', 'geoparks' and the extensive volunteer schemes that keep many environmental and development agencies staffed, often incorporate fieldwork as a core activity. Ever more landscapes are designated for protection and educational value and ever more people seek out forms of learning and leisure that interconnect and are active and outward looking. The quality of these fieldwork experiences varies enormously. However, current trends suggest that fieldwork is central to an expanding market in popular geography. Fieldwork is undergoing a new phase of mass participation.

Travelling and Knowing

Perhaps the most characteristic geographical practice is the journey. We all travel. But the task of filling in our mental picture of the world was once left to people for whom the journey could last many months or even years. From the early modern period onwards, such large-scale explorations were increasingly shaped by scientific considerations. The voyages of James Cook between 1768 and 1780 were defining moments of the new approach. The recording of flora and fauna by on-board naturalists, the collection of population statistics, and the charting of the transit of Venus, established Cook's explorations as one of the period's most impressive expressions of scientific endeavour. With Cook, as the historian of geography David Livingstone suggests, 'the tradition of scientific travel became firmly established'. Livingstone also points to another, simultaneous, function of European exploration,

> *[Cook's] circumnavigating instructions from the royal court included the specific objective of establishing British dominion on newly discovered soil and reporting on the natural resources, both organic and inorganic, that could be exploited by Great Britain.*[8]

Exploration is a political event. Indeed, because of its use by European powers to expand their empires, it has acquired unappealing connotations of racial arrogance and subjugation. To twenty-first century eyes explorers who seem least tainted by this imperial tradition appear more sympathetic. Thus, if Alexander von Humbolt's explorations (between 1799–1804) in South America, as well as his later trip to Central America (in 1829), retain an exemplary status it is partly because they strike us as relatively free from thoughts of conquest.

However, the act of travelling from a rich, powerful society to a poor and powerless one in order to bring back information will always provoke suspicions, both of exploitation and of what kind of information an 'outsider' can be expected to acquire. So what is the traveller to be? Three answers remain influential: a stranger; a participant; and a self-monitoring observer. The first answer remains the most provocative, for it is based on the idea that only an outsider can see through the fog of local prejudices. Montesquieu's *Persian Letters* (first published in 1721) adopted the figure of the stranger as a uniquely valuable and attentive individual, someone who is not blinded by prejudice. The letters of the title are an invented exchange of opinions between two Persian travellers in France, Rica and Usbek. 'You who, being a foreigner', a 'candid' French acquaintance suggests to Rica, 'want to know about things, and know them as they are'.[9] It is the foreigner who can see the wood from the trees: natives cannot truly understand their own society. Indeed, it is the very act of travelling, of being pulled from their roots, that enables Rica and Usbek access to the notion of cultural prejudice. This model suggests researchers should try to maintain their distance: that they should not put much faith in local 'gatekeepers' and informants. Rather they should try to obtain, as far as possible, a range of contacts, sources, and points of information. It is an approach that stands in stark contrast to the notion that the investigator should become fully 'involved'. The task of the researcher is, Malinowski argued, 'to put aside camera, notebook and pencil, and join in himself in what is going on'. Once done, he or she 'plunges into the lives of the natives'.[10] Immersion, or 'going native', seems to some the only way of knowing another culture. But it carries its own risks. The idea that one can unlearn one's own culture and 'plunge in' smacks of conceit. There is naïve headiness to such accounts which suggest that such leaps may tell us more about the diver than the pool. Certainly, the contemporary inclination is to show a certain scepticism towards claims of cultural immersion.

For the twenty-first century explorer the watchword is reflexivity. The field-worker encountering unfamiliar terrain is expected to monitor their own responses and their own capacity to misinterpret and to misuse the situation. It is indicative that our era's most influential cross-cultural researchers, such as James Clifford and Clifford Geertz, are best known for what they have written on other cross-cultural researchers.[11] This trend may also reflect the fact that the distinct and 'untouched tribal communities' that were once objects of Western fascination have disappeared. The present day researcher is likely to encounter subjects who are familiar with the expectations of the research process. The notion that knowledge requires that strangers encounter strangeness has had to give way to presumptions of familiarity.

Everyone an Explorer

On their doomed last haul across the Antarctic in 1912, Captain Robert Falcon Scott's party was carrying 35 pounds of geological specimens. They could have lightened their load. But then what would have been the point of going? Like Scott, the Royal Geographical Society, which had promoted his trip, saw exploration as a scientific, specialist undertaking. Both assumed that the era of exploration for its own sake was coming to an end.

And yet exploration has never just been about heroic deeds. We all have an instinctive curiosity about our world. For most people such feelings once had to be sated by tales from afar and limited forays from our own native spot. The modern era has changed that. A culture of mass exploration has emerged. It takes many forms, from package tourism to feats of endurance in challenging terrain. The world is understood to be diverse, exciting but also reachable. It is a world in which one can, for example, live in the USA and imagine India as exotic and distant but also fully expect to be able to be there in a matter of hours (and, moreover, back home, to find Indians living round the corner).

The immediacy and accessibility of the world's nations and environments allow individual biographies to be structured around moments of adventure and travel. Ask someone in the West what they have *done* over the past year and you are quite likely to hear a roll-call of flights and destinations. One area that has kept pace with the democratisation of exploration is travel writing. Internet 'blogs' and websites have turned travel documentation into something both ubiquitous and myriad. Exploration has emerged as one of the most creative and energetic forms of contemporary cultural expression.[12] This intellectual ambition is also reflected in the output of professional writers. For example, Bruce Chatwin's journey into southern Chile, narrated as *In Patagonia* (1977), explored an 'unfamiliar' landscape in order to discover what it is to be alive in an atomised and dislocated world. Chatwin renounced any claims to be offering descriptions of exotic natives and embraced the notion that travel reveals the traveller.

However, the most interesting contributors to contemporary travel literature show us that we may also be entering an era in which the most revealing journeys are local ones. Exploration is being 'brought home' into narratives of the unfamiliar and disorienting within our own immediate environment.[13] Iain Sinclair writes personal, dislocative, historical geographies of his travels amid the most prosaic landscapes. In *London Orbital* (2003) he travelled round the outer urban ring road of the city, glimpsing the hidden memories of mystics and utopian dreamers whose vestiges lie along the road's margins. Sinclair's concern is with the almost erased, the solitary gravestone, the unlikely place-name, the traces that rupture the ceaseless homogeneity of the commuter freeway.[14]

Sinclair understands that there is a slightly desperate edge to modern exploration. The local and the prosaic are now intriguing and novel but more distant horizons are clouded with cliché. The 'ultimate experiences' of a plane-hopping population are in danger of appearing shrink-wrapped and meaningless. Why do I say this? Perhaps it is because I have just been browsing *Unforgettable Journeys to Take Before You Die* (2003), not long after leafing through *1000 Places to See Before you Die* (2003), followed by *Unforgettable Places to* See *Before you Die* (2004).[15] And all I can think about now is death.

To Connect

We have trouble making connections. It is not the modern way. Ours is an era of specialisation. Yet geography's pre-modern, holistic roots contain considerable wisdom. It may not be the modern way but unless we take an integrated view of the relationship between human activity and nature the world will become uninhabitable.

The towering figures of nineteenth century geography – Humbolt, Ritter, Mackinder – offered *connection* as the core technique of the discipline. Indeed, they offered it as a model for all modern science. For Carl Ritter, geography,

> *aims at nothing less than to embrace the most complete and most cosmical view of the Earth; to sum up and organize into a beautiful unity all that we know of the globe.*[16]

However, Halford Mackinder knew this to be an increasingly unwelcome message. His plea, from 1887, for geography to be a 'bridge' between the natural and social sciences has a pleading tone:

> *The more we specialise the more room and the more necessity is there for students whose constant aim it shall be to bring out the relations of the special subjects. One of the greatest of all gaps lies between the natural sciences and the study of humanity. It is the duty of the geographer to build one bridge over an abyss which in the opinion of many is upsetting the equilibrium of our culture.*[17]

Today's environmental crises mean that Mackinder's words appear highly pertinent. We know that geography's integrating approach is a necessary commitment. Unfortunately, the lack of 'equilibrium' that Mackinder warned of affects many recent undergraduate textbooks on geographical methods, which often presume that human and physical geography are entirely discrete traditions.[18]

But how should integration proceed? Thankfully attempts to bridge the human and natural sciences are gathering momentum across many different disciplines. For example, within the relatively new fields of ecology, environmental studies and environmental management, this concern is to the fore. Connecting the human and natural sciences can take many paths. However, three basic steps need to be considered.

1 *Identification of parts.* In order to build up an integrated analysis we need to start by breaking reality down into fragments. These components can then be put back together again, in order to offer us a model of the whole. Let us take the example of desertification in Northern Kenya. Figure 4.1 shows a network of factors leading to this outcome.[19] It does not show the significance or even the form of the relationship between its constituent parts. What it does do is break down a complex system of human and natural components, picking out and classifying the significant factors. The focus in this model is upon the twentieth century and upon how the actions of governments have an impact upon over-grazing. This helps

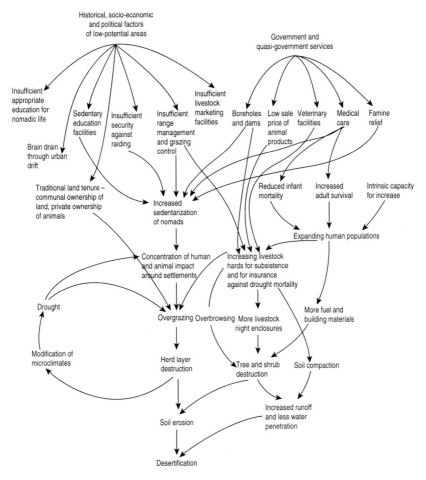

Figure 4.1 A network schema of some of the factors in desert encroachment in Northern Kenya

Source: Lamprey, 1981

explain the categories we see at the top of the diagram. There are 31 interconnected outcomes, each of which needs to be defined and justified. There is much room for oversight in such models. One looks at them and begins to think of other factors, other relationships. But this is part of their function. They provide us with a map of a complex whole, a tool to guide us through an interconnected set of problems.

2 *Defining period and scale.* As well as identifying components we need to think about their spatial and temporal parameters (the where and when). The example presented in Figure 4.1 addresses Northern Kenya (the where) in the twentieth century (the when). At a much wider scale, James Lovelock's model of self-regulating integration addresses the whole earth since its formation 4.5 billion years ago.[20] Any attempt to demarcate a discreet region or period of time will always contain a degree of arbitrariness. There are no truly natural regions or blocks of time, nor places and times which are not influenced by other regions and periods. However, as Richard Hartshorne advises:

> *The conclusion that regions are not definite concrete objects but merely arbitrary divisions of the earth surface made by the student does not cast out the problem of dividing the world, or any large part of it, into regions nor reduce to unimportance the basis for such a division. It is important to find the most intelligent and useful method, or methods, of dividing the world into regions.*[21]

The contemporary trend is to admit to these difficulties. There is also a current willingness to work with multiple scales. Multiscalar approaches encourage researchers to frame their topics within a diversity of spatial contexts.[22] Thus, for example, research on environmental change may need to simultaneously engage changes from the largest scale – the earth in the solar system – right down to micro-climates and micro-ecologies.

3 *Translating social into natural values and vice versa.* Social and natural scientists suffer from a problem of translation. Even when they understand each other's words, the meaning and the motivation often get lost. One of the obvious ways translation can take place is by reference to the dominant way things are given meaning in a capitalist society, that is by reference to *cost*.

A price tag can be put not only on environmental damage – translated in policy terms into the principle of 'polluter pays' – but also on other consequences of environmental change. Two other methods of translating scientific into social meaning concern aesthetics and risk. The perception of landscape attractiveness and the dangers posed by earthquakes both illustrate how natural forms can be translated into social values. 'Thus the gap between natural scientific facts and socio-economic aspects' Olaf Bastien concludes 'can be bridged'.[23] Of course, this is a one-way translation, from nature to society. The representation of human values as driven by evolution, and of human behaviour being, if only partly, genetically determined, suggests pathways for how translation in the opposite direction can be achieved. However, the problem of reification – of taking social categories and falsely claiming them as natural – should not be underestimated. Indeed, if an undiscriminating holism encouraged a reversion to treating controversial social categories, such as 'race', as natural entities it would soon become associated with an unhelpful politicisation of both natural and social science.[24] Knowing what not to 'translate' can be just as important than knowing what to.

To Map

Peter Haggett calls geography 'the art of the mappable'.[25] Richard Hartshorne claims that 'a simple test of the geographical quality of any study is whether it can be studied fundamentally by maps'.[26] These are strong claims. Too strong. The map is more than a useful tool for geography. It is the distinctive visual expression of the geographical imagination. But to root geography in mapping is a suffocating fantasy: geography can no more be reduced to maps than history can to a list of dates.

Such zeal makes it harder to appreciate the map for what it is: the most mesmeric and unique creation of the geographical tradition. From the first known map, of Catal Hyuk in Turkey, which has been carbon-dated to about 6000 BC, to Google Earth, people have felt a need to organise and plot their world in graphic terms. Technical rigour is not always an indicator of a map's usefulness. Asking ordinary people to draw so-called 'mental maps', crude sketches of the streets and places of importance to them, can be an invaluable aid in working out how they use their environment.[27] Nevertheless, there are good reasons for narrating the story of mapping in terms of the rise of technique. Mapping in Europe was advanced considerably by the use of the compass (with local maps adopting the convention of putting north at the top from the late fifteenth century) and the introduction of triangulation. Triangulation, first described in 1533 by Gemma Frisius, enables the measurement of straight lines over a curved surface (i.e., the earth). If one can measure a baseline, then the angles and distance to a third point, visible from both ends of one's baseline, can be gauged using a theodolite. In Britain systematic mapping began in 1791 with the trigonometric surveys of the Board of Ordnance. In the first decades of its existence the Board of Ordnance was required to provide information that could be used to prepare for an invasion by French revolutionary forces. Among the Ordnance Survey's first maps were the two likely invasion sites, one inch to a mile maps of Kent (published in 1801) and the South West (issued 1810). The Victorian apotheosis of triganometry came with the Great Trigonometric Survey of India, established in 1817 and completed in 1890.

Other early technical advances were the introduction of contour lines (first used by the Dutch engineer Nicholas Cruquius in 1728) and the admission of *terra incognito*. The latter may not sound particularly innovative. Swift's parody of the previous European custom helps us to grasp its novelty:

So geographers, in Afric-maps
With savage-pictures fill their gaps
And o'er uninhabitable downs
Place elephants for want of towns.[28]

When, in the late eighteenth century, the French cartographer Jean-Baptiste d'Anville began leaving blank spaces for those areas of his maps for which he did not have accurate, verifiable information, he was escaping millennia of mapping practice. No longer would hearsay be relied on to ensure a 'filled-in' portrayal of the world. Admitting the limits of knowledge is one of the hallmarks of scientific endeavour.

Map historian Jeremy Black argues that '[t]he world map is the most important continual preoccupation of cartography'.[29] Mapping the globe has, conventionally, meant taking something spherical and representing it as a continuous image on a flat surface. Try it with any orange and you'll end up with a rag of rips and gashes. Gerardus Mercator solved the problem in 1569 by imagining the earth was a cylinder, with the poles having the same circumference as the equator. If you do this then not only do you get a flat map (by unrolling the cylinder) but, even more importantly for Mercator, you get a map with a reliable navigational function. Mercator's projection keeps angles, and hence bearings, uniform. But it is also an ideological tool. It expands the size of the temperate zones and places Europe centre stage, at the top and centre of the world.

Although numerous attempts have been made to correct the distortions of Mercator's projection, the mapping of the world remains contested territory. Seeming advances, such as Arno Peters's map in 1973, which offered an 'equal-areas' projection in which land masses are presented in their correct proportions, can carry their own ideological baggage. Peters's pro-Third World disposition led him to solve the problem of mapping the world by contracting latitudes towards the poles. Thus the lengths of landmasses in the tropics are stretched. Mercator's Eurocentric map and Peters's Third Worldist map are two extremes between which cartographers have devised a number of less tendentious models. However, finding a representation of the world that does not seem to be making an argument about some bits being more important than others remains an elusive goal.

Today, mapping is no longer confined by the limitations of having to represent the world on paper. The development of computer and satellite mapping techniques has changed how we use maps and how we look at mapping. Such technologies have massively expanded the role of mapping across policy making, scientific research and everyday life. The first satellite photographs of Earth were made in 1960. However, it was in 1972, when the first of four LANDSAT satellites was launched, that the era of satellite mapping really began. The possibility of possessing up-to-the-minute maps of everything from the retreat of forest cover to local road networks, combined with advances in the computer manipulation of spatial data, has turned mapping into a key tool for the twenty-first century. As Virginia Gewin argues,

> *geospatial technologies have changed the face of geography ... by combining layers of spatially referenced data with remotely sensed aerial or satellite images, high-tech geographers have turned computer mapping into a powerful decision making tool.*[30]

Over the past few years we have come to expect the world to be mappable at numerous scales and in numerous ways. Cartographic Data Visualizers offer the possibility of taking any data set (number of sports shops, numbers of votes cast for one political party, etc.) and having the information presented to us as a map. Moreover, the Internet has seen a revolution in people's access to mapping techniques. The satellite and aerial data collected for Google Earth promise to allow anyone, anywhere, a God-like ability to see everything (although, at the moment, the images are up to three years old). Reflecting the nature of modern geographical consciousness, the start image for *Google Earth* is the whole earth, the blue marble. From here one zooms in, down to a level of detail that allows people in some parts of the world to pick out their own parked cars. Like other world maps, *Google Earth* is a political event. The further one ranges from the USA and other Western countries the more likely one is to find that one is dropping into unresolved mud. Type in 'Birmingham' or 'St Petersburg' and you will be led to cities in the United States. The controversies about the content and use of *Google Earth*, as well as for rival visualisation systems, have only just begun. A new era of map making is unfolding.

To Engage

Geography pushes its students out of the classroom and into the streets. This imperative implies another, that geography encourages a concerned and active disposition towards the world. The two main themes of contemporary geography, international knowledge and the environment, provide fertile ground for pro-poor and environmental activism. This set of associations has not been lost on some critics. One British journalist writing in February 2003 wondered

> *Are geography classrooms places where students are now taught to bow before the alter of environmentalism, while learning that multi-nationals and Western governments are the devil incarnate?*[31]

This provocation is a neat expression of the cynicism likely to confront geographers' attempts to be more than just passive observers. By having the world – its environments, nations and peoples – as its arena of concern, geography encourages an awkward, restless need to be involved, to do something more than study and describe. Lesley Cormack, writing about seventeenth century geography, helps us understand that this resolve to push at the boundaries of scholasticism is not new.

> *Geographers, who were interested in increasing and developing their knowledge of the world, had to be theoretical, practical, and political. Geographers then were neither scholars nor craftsmen, nor even states-men, but rather were a combination of all three ... Geography thus stands as a model for the type of investigation which allowed its seventeenth century practitioners to cast off the binds of scholasticism and to enter a new, engaged era.*[32]

Although it has a considerable historical pedigree, geography's engaged attitude fits well with the contemporary fashion for highlighting the value of *relevant*, *useful* and *accountable* forms of research activity. More specifically, it dovetails with the concern expressed by organisations, from the World Bank to most NGOs, for 'social involvement' and 'community consultation'. 'Having consulted widely' is a familar phrase today, often followed by the claim that communities must be given a sense of 'ownership' over the policies and decisions that affect them. Studies of international and environmental change increasingly cast community participation and community knowledge as both key methodologies and necessary resources. In *Involving the Community*, Guy Bessette identifies the issue of communication as central to the success of 'participatory research'.[33] His communications-based model is designed for researchers working in developing countries, but could be applied anywhere. His approach has ten stages (Figure 4.2), which take us through from initial contact to identifying how methods and results will be disseminated.

One of the areas where a participatory approach is being employed is within environmental research that draws on so-called 'indigenous knowledge'. The idea that indigenous people have reserves of knowledge about how their environment can be sustainably managed may be exemplified by reference to Trevor Wickham's study (conducted in the early 1990s) of an isolated village in Bali, Indonesia.[34] The villagers were asked to complete village maps showing topography and land use. They also took researchers on walking tours of the village and its surrounding land. The villagers identified 146 tree types and eight soil types. This compares to the 16 tree types and one soil type known to 'experts'. The researchers were also able to learn about a number of sustainable practices deployed by the villagers.

> To limit soil erosion, farmers used a combination of techniques – they maintained vegetation cover, cut terraces, practised strip cultivation, and planted perennials and annuals together. They used green manures and mulching for soil-fertility management. The farmers' techniques for managing weeds included multiple cropping with fallow, mulching, and selective weeding.[35]

Participatory research and working with indigenous knowledge are now mainstream practice. However, geography also nurtures a number of deliberately challenging activist traditions. Some of

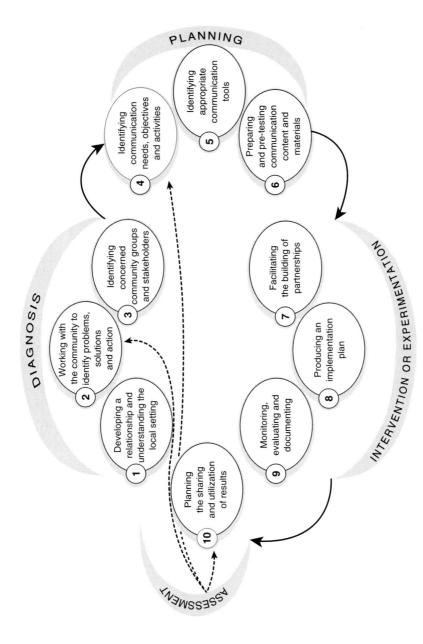

Figure 4.2 The participatory development communication model

Source: Bessette, 2004

these reflect the issues discussed in Chapter 3 and concern urban development, migration and anti-racist campaigning.[36] However, the principal forms of activism mirror geography's modern intellectual axes: they transform the researcher of the environment into the *environmentalist* and the international specialist into the *pro-poor campaigner*. By number, charities and other voluntary organisations representing these two issues form by far the largest agitational movements on the planet. The great majority of these organisations rely on voluntary donations in order to sustain both campaigning work and direct interventions, such as emergency relief or environmental protection.

One of the most extreme forms of environmentalist activism is represented by Earth First!. In *Ecodefense: A Field Guide to Monkeywrenching* (1987), founder member Dave Foreman set out the group's adherence to extra-legal direct action. His colleague Edward Abbey explained:

> *self-defence against attack is one of the basic laws not only of human society but of life itself, not only of human life but of all life. The American wilderness, what little remains, is now undergoing exactly such an assault.*[37]

The willingness to use 'direct action' finds echoes among activists campaigning on issues of global development, particularly among those who are part of the anti-capitalist and anti-globalisation movements.[38] The meetings of the World Trade Organisation and G8 are regular sites of protest. However, the high-profile activism of militants represents only a tiny part of the culture of mass protest that emerged in the mid-to-late twentieth century. International development and environmentalism are matters not merely of interest but of involvement for many millions of people today. As we shall see in the next chapter, over the last 50 years popular campaigns on poverty and debt relief, nature protection and climate change have sprung up, campaigns which are shaping contemporary political priorities. The core concerns of a geographical education have emerged as the core concerns of the global citizens of the twenty-first century.

Geography, Children and Freedom: A Plea

A study of geography's methods shows us that the geographical imagination is both a specialist and a popular enterprise. From the researcher of climate change pulling cores from the Greenland ice sheet to a bleary eyed gap-yearer finding their way to a taxi at Lima airport: geographical practice can be witnessed millions of different ways every day. It cannot be shut away in a university library.

The geographical imagination is insatiable. Even if geography were to be dropped from every school and university curriculum, some kind of illiterate geographical consciousness would blunder on.

Yet if any area of geography makes me worry, it is the area of practice. I worry because people are increasingly insulated from the natural world. Indeed, in many societies, we are actively *protected* from the world outside our doors. Writing of Australia, Karen Malone and Paul Tranter observe that

> *Spontaneous unregulated play in neighbourhood spaces, particularly in affluent areas of the cities, is increasingly becoming an activity of the past. Many children have lost access to traditional play environments, including streets and wild spaces, partly through parental fears about traffic danger, bullying and 'stranger danger', partly through the loss of natural spaces and partly through perceptions of what is best for children.*[39]

Geography requires freedom. In case this sounds too romantic let me clarify. Good geography, a geography able to explore, connect, map, and engage, requires freedom. Yet freedom is easily lost. The society of long-work hours, constant surveillance and materialist, isolated lives, is a society in which freedom is dying and with it the large horizons and outward disposition of geography. One of the most tragic aspects of this enclosing state of comfortable passivity is how it affects children. Increasingly, children are deprived of the freedom to explore, to roam and hide away out of reach of adults. The traffic and 'stranger danger' mean that the streets are too risky, so too the parks. And what really is the point of unsupervised play?

In a culture where children are expected to always be visible, to be meeting 'targets' and achieving 'outcomes', this question becomes harder to answer. Both work and play are now scripted as useful and productive 'opportunities'. For freedom and adventure read 'supervised outdoor learning experience' and 'professionally designed educational trip'. Children soon find that their only 'down-time', the only place where they are not being appraised, is when they are sat in front of a console or TV. In depriving children – and adults too – of environments in which they are free to explore and where risks can be taken, we are creating the conditions for the hollowing out of the geographical imagination.

5 ●● Institutionalising Geography

Introduction

What would we do without our councils, boards, committees, societies, associations and institutes? Ours is a highly organised and organising culture, in whose ocean of bureaucracy we all must navigate. Geography has been round long enough for numerous institutions to accrete around its name. Some are now thriving, some dying. It is also old enough for us to wonder at this process.

Geography is as much a pre-modern as a modern enterprise. Both its heady claim to world knowledge and its inclination to take the human and natural world as one, connected, whole, contain echoes of eras less form-filling and fragmentary than our own. Not unrelatedly, geography has a rich popular tradition, sprawling from travellers' tales to environmentalists' direct action. To professionalise geography, then, is no easy task.

This chapter charts two dilemmas, each of which goes to the heart of geography's modern story. The first is that, although geography *has* been institutionalised, there remains something 'uninstitutionalisable' about it. Indeed, to try and cage it is to set in motion an unending debate on geography's coherence as an institutional enterprise. It is also to witness an unstoppable swarm of sub-fields flying off geography (from geology to international studies) and setting up their own organisations. However, the second dilemma concerns the fact that institutionalisation is not simply a way of restricting and narrowing geographical knowledge. It is also a necessary step towards opening geography to a mass audience. The rise of school geography provided a geographical education to every citizen. Institutionalisation tries to corral geography. It is a continual, impossible, struggle. But, on balance, it seems likely that without this struggle, geographical knowledge would be far less accessible.

There are, then, some rich ironies in geography's institutional story. Few of them can be appreciated if one believes that the story of geography begins with its entry into universities and continues as a set of debates among academic geographers. 'A historian of geographic ideas', Clarence Glacken tells us, 'who stays within the limits of his discipline sips a thin gruel'.[1] Geography's borders are broad and permeable. If we address geography as a world discipline, then we must address its institutionalisation in terms of the institutionalisation of world knowledge. Moreover, the nature of geography also demands that we consider both *specialist* and *popular* institutions.

Specialist Institutions

Eratosthenes coined the word 'geography' over 2,000 years ago. Until the last 200 years or so its institutional forms were mostly associated with colonial endeavours and the diplomatic activities of courts and states. The first half of the nineteenth century saw a great expansion of imperial and commercial ambitions. This period also witnessed a flurry of new geographical organisations. Some had a rather dilettantish image, such as the Travellers' Club, founded in London in 1819. Entry to the club stipulated that would-be members had 'travelled out of the British isles to a distance of a least 500 miles from London in a straight line'.[2] More practical goals guided the formation of the Royal Geographical Society in 1830 (a year later the RGS incorporated the African Association, established 1788). The first issue of the RGS's journal proclaimed,

That a new and useful society might be formed, under the name of THE ROYAL GEOGRAPHICAL SOCIETY OF LONDON.

That the interest excited by this department of science is universally felt; that its advantages are of the first importance to mankind in general, and paramount to the welfare of a maritime nation like Great Britain, with its numerous and extensive foreign possessions.[3]

The nineteenth and early twentieth century saw the foundation of geographical societies, associations and institutes across Europe and North America. In Britain in 1893, five teachers formed The Geographical Association, which continues to promote the subject in British schools. The National Geographic Society came into being in 1888; its stated mission 'the increase and diffusion of *geographical* knowledge'. In 1904 the Association of American Geographers was established as a professional body for academic geographers. In 1933 the Institute of British Geographers (IBG) was set up to perform the same function for British academics (the RGS and IBG merged in 1995).

From the end of the eighteenth century we also begin to see areas traditionally considered to be part of geography begin to form distinct institutional structures. Thus, for example, the Geological Society was formed in London in 1807 and the Zoological Society in 1826. The study of particular regions of the world had already come to be demarcated in a similar fashion. The founding of the Asiatic Society in Calcutta in 1784 was an early and influential moment in this history. William Jones, the linguist who started the Asiatic Society, explained that its scope would be 'MAN and NATUR: whatever is performed by the one, or produced by the other'. The only intellectual limit he placed upon the Society was that it confine its enquiries within the 'geographical limits of Asia'.[4] In this way, 'area studies' was born. Where the 'Asianists' led, many others were to follow. 'Africanists', 'Sinologists' and 'Latin Americanists' developed their own societies and, eventually, academic departments.[5]

From the start of the twentieth century, other specialised themes in world knowledge that had once found a place under geography's inclusive umbrella, such as planning, environmental studies, anthropology and international politics, also began to emerge as distinct institutions. The requirements of the increasingly complex modern state demanded that this process often took place across a number of different sectors. Thus, in Britain, the Town and Country Planning Association (founded 1899), the Housing and Town Planning Act of 1909 and the establishment of university departments, such as the School of Planning at University College (in 1914), may be seen as forging an interconnected array of organisational structures. Later, a similar process may be witnessed in the area of environmental management and research. The flows of funding, information and personnel between academic departments in 'environmental science' and 'ecology' and what, in Britain, is the Government Department of Environment, Food and Rural Affairs (which partly funds the Environment Agency, formed in 1996) and numerous non state agencies, have created an inextricable institutional network.

How does geography fit into all these organisations? We see geography becoming more institutionalised but also becoming a victim of institutionalisation. The interest in and activity around geographical themes expanded. But, increasingly, these themes were treated as separate fields of inquiry and disconnected from each other. In many countries the only place where this has not occurred has been school education. Thus, geography at primary and secondary school level often provides a stepping stone to the pursuit of specialisms in university or professional life. But within school geography itself, the *world*, from geology to planning to development, continues to be addressed. In his account of school textbooks in the USA, John Nietz found that geography continued to thrive despite its status as 'the mother of many other subjects'.[6] The first American geography textbook for school use was Jedidiah Morse's *Geography Made Easy*, published in 1784. However, the dramatic rise of geography as a school subject had to wait till the expansion of mass education in the late nineteenth century. In Britain school geography expanded along with the British empire. The expectation that pupils should be able to recall the resources, communications, topographic features and ethnic groups of Britain's overseas possessions came to dominate the subject. In an educational climate inclined to repetitive exercises of 'read and remember', geography was crushed into factual dust. The late Victorian period represents a high water mark of this kind of geography. As J. R. Green observed with bitter disappointment, in his *A Short Geography of the British Isles* (published in 1879),

> *No drearier task can be set for the worst of criminals than that of studying a set of geographical textbooks such as children in our schools are doomed to use ... they are handbooks of mnemonics, but they are in no sense handbooks of geography.*[7]

Geography has had the life pummelled out of it by professional geographers on a number of occasions. Yet they never quite kill it. In the compendiums of world facts that lay, brick-like, on children's desks for so many years, the wondering, creative child's mind could still see glimmers of the strangeness and diversity of the world.

The rote learning of geographical facts was in keeping with Victorian educational methods. Its survival in school geography into our own time is less explicable. The notion, strongly prevalent when I was at school in the 1970s, that geography is a subject that rests on having a good selection of coloured pencils at hand in order to draw the outlines and map the main resources of randomly selected countries, continues to shape public attitudes to the discipline. It should not be forgotten, however, that some of the pioneers of school geography always had a much wider vision of the subject. For Thomas Huxley and many other inspiring teachers, school geography was always about understanding and engaging the world outside the window. Huxley illustrates the enthusiasm for world knowledge that, despite so many dull textbooks, has often brought together student and teacher. Speaking in 1867 (and using the German for 'geography', *'Erdkunde'*) he enthused,

> *If anyone who has had experience of the ways of young children will call to mind their questions, he will find that so far as they can be put into any scientific category, they come under the head of 'Erdkunde'. The child asks, 'What is the moon, and why does it shine?' 'What is this water, and where does it run?' 'What is the wind?' 'What makes this waves [sic] in the sea?' 'Where does this animal live, and what is the use of this plant?'.*[8]

The nineteenth century saw the institutionalisation of geography and some of its manifold sub-fields at a national level. However, in the twentieth century geography's new establishment increasingly had to fight for attention in the shadow of institutions that operated internationally. Indeed, it may be argued that these international institutions are far more in sympathy with geography's global claim. Certainly, by the end of the last century, the wider public and many teachers often looked for inspiration and guidance on environmental and international research and initiatives to pan-national organisations, most notably the United Nations. As this implies, although some new international institutions sought specifically to bring together professional geographers (notably the International Geographical Congresses, the first of which was held in Antwerp in 1871 and the International Geographical Union, founded 1922), they were dwarfed by the emergence of others whose activities address geography's traditional concerns but which were not bound by academic boundaries. Some of these institutions can trace a long history of international cooperation. For example, the World Meteorological Organisation was established in 1873 in Vienna (as the International Meteorological Organisation). It gained status as a specialised agency of the United Nations in 1951. The United Nations has dozens of specialised agencies, most of which have an interest in either international or environmental research. The two largest are the World Health Organisation (1948), and the United Nations Educational, Scientific and Cultural Organisation (UNESCO, founded in 1940). These agencies support and undertake research and feed their findings into international policy making. Important outcomes have been the Declarations of Stockholm (1972) and Rio (1992) on Environment and Development and the Kyoto Protocol (1997: by 2006 this had been signed by 166 countries) on reductions in 'greenhouse gases'. The UN also supports complementary research networks. One example is the International Council for Science (founded 1931). Funded by UNESCO and member subscriptions, the Council currently sponsors four major global change programmes (the International Geosphere– Biosphere Programme; the International Human Dimensions Programme on Global Environmental Change;

World Climate Research Programme and DIVERSITAS, a programme of biodiversity science).

All this activity might lead us to wonder about the function and role of academic geography. As we shall see, although often cast as the centre of geography's institutional life, academic geography has found it difficult to balance the wide-ranging nature of geography's subject matter with the pursuit of intellectual and specialist status.

Academic Geographies

In 1871 the Royal Geographical Society offered the following justification for geography's acceptance as a university discipline:

> *We would point out the special importance of geography to Englishmen in the present age. The possession of great and widely scattered dependencies, the unprecedented extension of our commercial interests, the increased freedom of intercourse and closeness of connection established by means of the steamship and the telegraph, between our country and all parts of the world, the progress of emigration binding us by ties of blood relationship to so many distant communities – all these are circumstances which vastly enhance the value of geographical knowledge.*[9]

The practical and ideological significance of geography in an age of empire remained its greatest asset well into the twentieth century. The trend witnessed in Europe and the USA was for the establishment of geography lectureships and Chairs in the nineteenth century to be followed, in the first three decades of the twentieth, by the formation of university departments. In Germany the earliest appointment was a Chair in geography at Berlin, held by Carl Ritter from 1820. Subsequent Chairs in geography followed at Leipzig in 1871 and Halle in 1873. By 1914 there were 23 professors of geography in Germany. In Britain, Halford Mackinder was appointed to a Readership in geography at Oxford in 1887. The following year saw the creation of a geography lectureship at Cambridge.

One scenario is that, after a burst of Victorian eagerness, academic geography began to disappear as a discipline in the wake of the separate institutionalisation both of its sub-fields (geology, planning, regional studies and so on) and its international ambitions. To some extent this is, indeed, what happened. The mid-twentieth century saw both the expansion of these specialist fields and the contraction of academic geography in many parts of the world. Explaining the closure of the Geography Department at Harvard in 1948, the University's President announced that 'Geography is not a university discipline'.[10] Not unrelatedly, by the mid-twentieth century, the pressure to turn academic geography into a 'normal looking', entirely modern, specialist pursuit became considerable. Geography's history is still often told as a struggle to retain respectable, university status.[11]

Yet the attempts of academic geographers to identify geography as a specialist activity have never been entirely convincing. In fact, such efforts have sometimes undermined the discipline. For the pursuit of specialist status has, on a number of occasions, encouraged academic geographers to latch on to voguish core theories for the discipline. Around the turn of the nineteenth century, theories of evolution were taken up enthusiastically by academic geographers. Geography has traditionally been interested in the connections and 'bridges' between humans and nature. Yet, at the time of geography's entry into academia, this interest was often garbed in the – then modish – notion that both natural and social phenomena were subject to the law of the 'survival of the fittest'. In *Political Geography* (1897), Friedrich Ratzel applied this idea to geopolitics, giving us the idea that states are engaged in a battle for 'living space', or *Lebensraum*. Successful states required what Ratzel called large space, or *Grossraum*. In physical geography, William Morris Davies applied evolutionary ideas to the landscape. The cycle of erosion, he explained in 1885, was a form of 'inorganic natural selection'. For Davies it was evolution, the 'evolution of land forms' and the 'evolution of living forms', that constituted geography's bridge between the human and the natural.[12]

The theory of evolution seemed a secure path towards establishing geography's intellectual credentials. Yet questions soon began to be asked about the application of evolution to such disparate processes as erosion and warfare. Moreover, in the context of the rise of Nazi Germany and fascist Italy, the politics of aligning geography's mission with the predatory ambitions of aggressive states seeking *Grossraum* became repugnant. However, academic geography in the first half of the twentieth century had other intellectual resources. More specifically, it also offered itself as a 'normal-looking' academic specialism by reference to another major theme, namely the region. The idea that geography is the study of regions has deep roots in the history of geography. And while the idea that geography could be defined around the application of the theory of evolution had a rather short shelf-life, geography's regional ambition remained influential throughout the last century and maintain an important role in the academic discipline today.

The conviction that regions are at the centre of geographical enquiry has been expressed in many different ways. Fenneman, in 1919, argued that the concept of the region was 'a safeguard against absorption by other sciences'.[13] Isaiah Bowman echoed the orthodoxy, noting that geography's 'main purpose is regional analysis and if possible correlation'.[14] Although modern regional study may be rooted in Humbolt and Ritter's synthetic approach, it can also be traced to the French geographer, Paul Vidal de la Blache. In 1922 Vidal de la Blache offered an influential triptych of analytic tools to prise open the region: the *milieu* (described as what holds the region together, what makes it an homogeneous form); the *genre de vie* (the way of life exhibited by people in the region); and *circulation* (the nature of interchange between regions).[15] Writing in 1990, Peter Haggett went so far as to suggest that,

> *The central role of the regions has been so widely accepted within the geographical discipline that asking a geographer why he studies regions is like asking a Christian why he studies the Gospels.[16]*

The idea that geography is the study of regions enabled the discipline to maintain its concern with connecting the human and physical and assert itself as a distinct specialism.

However, the claim that regional study forms the *essence* of geography is difficult to sustain. What, after all, defines a region? Critics in academic geography, notably Kimble and Paterson, questioned both the intellectual clarity and role of the concept.[17] Recent years have seen the development of more sophisticated versions of regional studies. Today one is likely to hear geographers talk of 'multi-scalar' analysis and of regions as constructed through social practices such as institutionalisation (for example, the formation of regional government).[18] Yet the proponents of such methods rarely argue that regional study is the defining attribute of geography. Regional studies has become just another of geography's many sub-fields, with its own journals and, increasingly, a separate institutional structure.

In summary, the two main attempts by academic geographers in the early twentieth century to claim that geography can be defined as a specialist activity had limited success. Broader definitions of the academic discipline were required. The most famous academic definition of geography comes from Richard Hartshone. In *Perspectives on the Nature of Geography* (1959) he wrote that,

> geography is concerned to provide accurate, orderly, and rational description and interpretation of the variable character of the Earth's surface.[19]

It is, as Peter Haggett notes, the 'best known and most widely used formal definition' of the discipline.[20] It may be formal but it is also nebulous. After all, who ever thought they would set out to 'provide' the inaccurate, disorderly, and irrational? Or, indeed, that one could forgo 'description and interpretation'? The 'variable character' of the 'Earth's surface' seems to promise more. But the term 'surface' is misleading, since a lot of geography is concerned with climate or subsurface processes. One might admit 'variable character' if it did not seem like an attempt to turn the obvious into a scholastic achievement (would an historian feel the need to say history concerned the *variable character* of the past?). What are we left with? One thing, the one thing that lives and breathes and excites under all this tautology. The Earth. We recognise it as geography's province at once. The idea that geography is the world discipline threads its way through many verbose academic definitions like a strand of gold.

However, in the latter half of the twentieth century, academic geographers continued their search for specialist status. Moreover, they also began to loose faith in geography's integrative ambitions. The discipline began to fragment internally. Thus, academic geographers moved apart intellectually. Today, human and physical geographers are distinct academic communities.[21] This divorce represented another phase in the search for specialist status. Many physical geographers cast themselves as natural – as opposed to social – scientists. This purified ambition allows questions of environmental policy and human-natural systems to be framed as secondary. In human geography, the defining concepts that emerged were 'space' and the 'spatial'. Human geographers began to argue that geography (by which they increasingly meant human geography) could be defined as the study of spatial processes. Post World War II, the phrase 'spatial *analysis*' became associated with quantitative geographers. Today, spatial analysis is allied to spatial statistics and computer-generated mapping techniques, such as GIS. However, the notion that geography is about society *in space* is deployed throughout human geography. A wide variety of human geographers have used the concept to explain the discipline's specialist status. For Hartshorne, geography 'studies the spatial sections of the earth's surface, of the world'.[22] Marxist geographers argued that they were 'adding space' to Marx.[23] For post-modern geographers too, such as Michael Dear and Steven Flusty, geography's contribution is the 'spatial'.

> *Human geography is that part of social theory concerned to explain the spatial patterns and process that enable and constrain the structures and actions of everyday life.*[24]

However, outside of quantitative spatial analysis (where spatiality is, indeed, analysed), human geographers have shown themselves far more interested in understanding real world issues than abstract questions of space. Human geography journals carry papers on globalisation, migration, regional economics and a myriad of other topics that reflect the broad remit of the discipline. 'Space' and the 'spatial' are often invoked but rarely provide the focus of either the argument or the method. For the most part the language of 'space' is a rhetorical device, a red-herring dangled by professional geographers still hankering to be part of a 'normal-looking' academic specialism.

The most ambitious argument 'for space' comes in a recent book by Doreen Massey, titled *For Space* (2005). For Massey to consider space is to consider diversity, 'the sphere of openended configurations within multiplicities'.[25] Hence to turn towards the spatial is to move away from linear, purely historical arguments, and embrace the simultaneous co-existence of multiple social forms. This vision of space '*as* co-eval [co-evolving] becomings'[26] reflects the emergence across the social sciences and humanities of an interest in challenging Western ethnocentrism, more specifically the conflation of modernity with Western civilisation.[27] However, this latter project has been well underway within area studies for the past two decades. The concepts of multiple modernity and multiculturalism have helped frame it. Being 'for space' may add to this endeavour but is unlikely to shape it. Space is a purely abstract category. To fill it with social or political content, in the way Massey wants, is to conflate philosophy with social theory. As this implies, 'space' could be claimed by any number of projects: multiculturalists might be 'for space' but so might imperialists, or indeed any other project that demands the imaginative or actual seizure of diverse territories.

In the wake of the difficulty of finding a plausible specialism, academic geographers have begun to look favourably upon the idea that geography cannot and should not be defined; that it has no core and that it is inherently and characteristically *inter*disciplinary. Indeed, it has been noticed that interdisciplinarity does not go far enough, for it implies that disciplines must be sustained (in order that one may be 'between' them). Hence, bolder voices have endorsed a *post*-disciplinary identity.[28] The vision of geography as a meeting place of scholars who are far more interested in solving substantive research questions than worrying about the policing of anachronistic intellectual barriers has a certain appeal. It allows a virtue to be made out of an inability and unwillingness to tell the public what geography is. Georges Benko and Ulf Strohmayer contend that the current 'recognition of geography as an intellectually vibrant discipline' has been 'hindered neither by methodological eclecticism nor by the lack of a unified and recognisable focus'.[29]

My argument in this book implies that 'eclecticism' and a clear idea of what geography is can co-exist. More than this; I have tried to show that geography cannot be understood as just another academic discipline: that to defend and assert geography is to defend and assert the possibility of world knowledge. Academic geographers sometimes seem to delight in lashing out against geography. Geography is cast as an oppressive father, restricting the freedom of its progency. But this anger is misdirected. For it is not geography but the bureaucratic forms and mechanisms that surround and succour modern education systems that hinder the ability to connect, challenge and synthesise. And since the modern era is addicted to institutionalisation, with the endless flowering of regulatory agencies, hopes of a less intellectually policed future are likely to be disappointed. Indeed, over the past 20 years, new administrative structures that shape and even dictate what is to be taught and studied have blossomed in drab profusion. These structures, which range from National Curriculums to research funding 'priorities' and 'assessment exercises', reflect centralised, politically derived decisions about what kinds of activity are considered to be useful for social cohesion and economic growth. The culture of continual surveillance and auditing that has emerged establishes bureaucratic systems as the prime audience for scholarly and educational activity. In many countries today teaching is, in large part, designed for inspectors and research is written for assessment exercises.

It may be argued that geography's intellectual heterogeneity means that it is highly adaptable and, hence, suited to a politicised educational regime in which teaching and research staff have to respond quickly to the changing priorities of centralised bureaucracies. But geography's pre-modern legacy, its wide horizons and popular constituency, make it an elusive target for bureaucrats. A tendency, already apparent in some countries, for geographical education to be framed as a series of 'transferable skills', from numeracy to time-management, helps manage this most awkward of disciplines. For a field that has not been able to arrive at a satisfactory academic identity, a 'skills based' agenda offers temporary balm. But the problem of how to audit a subject as wide-ranging and institutionally unkempt as geography is likely to persist.

The further academic geography becomes enmeshed in educational and state bureaucracies the more apparent it is that the geographical imagination must be kept alive in other settings. Earlier I discussed a number of specialised institutional environments where world knowledge continues to be embraced. However, some of the most important and energetic institutions that are carrying forward the geographical ambitions of international and environmental study and practice are to be found in the realm of popular campaigning and communication.

Geography's Popular Institutions

It is only relatively recently that geographical institutions have become part and parcel of ordinary lives. One way this occurred was through the development of a mass audience for maps and world and national guides. One of the earliest and most influential of such commercial geography publishers was Justus Perthes, founded in Gotha in 1785. As well as maps, Perthes issued world almanacs and geographical year books. Such enterprises are just one small element of what we might call the 'geography media'. Over the last 200 years there has developed an insatiable public appetite for stories about the natural and human landscapes of the world. Geography's specialist institutions have often been at the centre of popular geography. In the mid-to-late nineteenth century, the public lectures of returning explorers regularly turned the Royal Geographical Society into 'a theatre of national suspense'.[30] However, by far the most influential of such institutional border-crossings are associated with the National Geographic Society. Today the National Geographic's various magazines and broadcast channels attract some 200 million readers and watchers. The glossy *National Geographic* magazine is published in 31 languages.

The National Geographic Society has branched out to cater for the increasing desire and, for many, ability to fulfil their curiosity through international travel. Its more recent puplishing ventures include *National Geographic Traveller* (founded 1984) and *National Geographic Adventure* (founded 1999). The National Geographic Channel, launched internationally in 1997, competes with other geography media providers, which in the English speaking world include the Discovery Channel and the Travel Channel. However, unlike most of its rivals, the National Geographic Society is a non-profit organisation. This allows it to continue to develop and mix popularist fare with research-based educational material. One of its largest recent ventures has been the National Geographic Genographic Project. Based on 100,000 DNA samples, and launched in 2005, this research aims to provide a detailed global history of human migration. The National Geographic has also worked to raise the profile of geography in the USA. It has helped establish a National Geography week in the nation's calendar and institute Geographic Alliances that bring together geography teachers, both nationally and on a state wide basis.

The National Geographic Society offers 'the world and all that's in it' to a mass audience. If we accept this interpretation of geography then it follows that geography's popular institutions are not limited to niche providers but embrace the full variety of institutions that organise and supply world knowledge. I will pick out and very briefly exemplify three of the most important: development and environmental campaigning, travel, and international news.

Development and environmental campaigning   If you are looking for institutions that are committed to the holistic integration of human and natural knowledge, as well as to environmental and international knowledge, then you cannot ignore campaigning groups. Indeed, the process whereby environmentalist and pro-poor concerns are institutionalised and handed over to professionals, such as academics and policy makers, is keenly observed amongst many activists. More specifically, those who feel that these commitments require passion and public participation often view the input of professionals with suspicion. Lorna Salzman, former staff member of Friends of the Earth, argues that,

> *The environmentalists are now entrenched. They're professionals. They're not in it for a cause – they're in it because it's a 'public interest', 'highfalutin' job. You've got a new group of professionals, bureaucratic professionals – all right, yuppies – and the idealist ones aren't there anymore; they've gone.*[31]

Many contemporary campaigning groups began as national bodies but have evolved into international organisations. Thus, the environmentalist group, Friends of the Earth, which was founded in the USA in 1969, became an international network in 1971 and now operates in 70 countries. Similarly, Greenpeace, founded in Vancouver in 1971, today has 41 national offices. Greenpeace's administrative and policy-making body 'Greenpeace International' was founded in 1979. In the case of Greenpeace the transition to an international central body augured a move away from a loose federal structure and towards a more centralised organisation.

Campaigning groups increasingly wish to engage transnational companies and international agencies, such as the United Nations. In this context, a professional and globally integrated approach to campaigning is often seen as an inevitable step. This is also true of development and aid campaigns. Thus for example, Oxfam, founded in Oxford in 1942, today has 13 national branches, all part of Oxfam International, founded in 1995. Oxfam is also a member of the One World Network, an Internet-based organisation founded in 1995, to provide a common communications portal for development campaigners. Another British-based example is War on Want, which was founded in 1951, the same year as International Workers Aid, the NGO alliance it was later to join (now called Solidar). The tendency across these campaigning groups is towards further international integration.[32] In a globalising world, purely national activity is increasingly cast as ineffective.

Travel   The geographical imagination ranges across many institutional forms, the scholarly and educational as well as the popular and pleasure driven (tourism is also discussed in Chapter 4). Often these motives connect. Holidays can be profound learning experiences and research field trips are often fuelled by the desire to leave office life many miles behind. The possibility of such connections is to the fore in many popular geographical institutions, whether it is the travel company promising to take visitors to 'authentic sites' or the TV documentary promising to change your perspective on a foreign country. In most countries the opportunity for ordinary people to travel for pleasure emerged from changes in employment law. Thus, for example, in Britain, statutory free time began with the Bank Holiday Act of 1871. With the help of new travel technologies, notably the railway, then the car and the aeroplane, such entitlements became associated with travel to ever more distant destinations. They also became increasingly mediated by commercial institutions.[33] Holidays began to be sold as 'packages' from the mid-nineteenth century. This was an idea first tried out by the tour operator Thomas Cook in 1841, when he sold an inclusive package (travel and admission) from Leicester, to a temperance rally in Loughborough. Another trailblazer was the *Michelin Guide*. In 1900 Michelin published a handbook for drivers touring France, which has since become a model for numerous other guides. Michelin still employs its original, highly influential, star rating for 'sites': three stars 'worth the trip'; two stars 'worth a detour'; one star 'interesting'. It is an efficient, accessible, utterly reductive way of looking at the world. The institutions of travel shape not just how and when we travel but also what we see and feel.

International news   One of the greatest changes that has occurred to our understanding of geography concerns how we acquire news about the world. As with travel, what was once the preserve of a small elite is now something widely available in many different forms. The expectation that we should all be informed about our world has become realistic only because of the formation of a mass media. The history of the mass media shows a continuous effort not only to provide world news, but to claim such provision as its *raison d'être*. Despite the parochial content of a lot of national daily news bulletins, global news is delivered by a growing range of institutions.[34] One of the oldest is Reuters. Paul Reuter opened an international news agency in London office in 1851, acquiring offices in East Asia in 1872 and South America in 1874. In 1923 Reuters pioneered the use of radio to transmit news internationally. Another well known global news institution is the government funded British Broadcasting Corporation (BBC). Its World Service (founded as the BBC Empire service in 1932) broadcasts radio programmes in a number of languages. (The selection of languages to broadcast offers a revealing picture of which countries the British government was seeking to influence: Polish was broadcast from 1945–2005, Japanese from 1948–1992.) Today, the BBC broadcasts television news around the world and is seeking to compete with other international media companies, both Western but also increasingly non-Western in origin, for satellite, digital, Internet, and terrestrial markets.

Access to global news has never been more readily available. The geographical imagination of the twenty-first century grasps the world in ways that are more immediate and institutionally diverse than ever before. It is more debatable, however, that people are better informed than in the last century or, indeed, that the diversity of global news reflects a greater diversity of opinion and content. Global media companies both reflect and are part of an ongoing economic and cultural globalisation: news services across the world look and sound very similar. Finally, a polemical point. For I would also suggest that today the world is often presented as an entertaining spectacle for passive and disconnected individuals. We have become mere image consumers who, even though we may spend many hours tuned into some screen or other, often appear to be unengaged either with each other or the outside world.

Conclusion

I've been an academic geographer ever since I started a university degree in the subject in 1983. In a way, it's all I know. But the words 'academic geographer' still sound odd to me. Observing the academic institutionalisation of geography, the novelist and traveller Joseph Conrad acidly remarked that it was producing,

persons of no romantic sense for the real, ignorant of the great possi-
bilities of active life; with no desire for struggle, no notion of the wide
spaces of the world – mere bored professors ... their geography very
much like themselves, a bloodless thing with dry skin covering a repul-
sive armature of uninteresting bones.[35]

There is more than a touch of prow posturing here. Still, Conrad was surely right to kick out against the suffocation of geography's will to meet the world; to rail against the enclosure of something essentially outward looking. If geography is to have a place in schools and universities it is necessary that its awkward nature is recognised and valued. Geography can be learned about in classrooms but it also takes us outside, to places where we can see the classroom as the little box it is.

The institutionalisation of geography is part and parcel of a wider democratisation of knowledge that has occurred over the past 200 years. What was once thought of as only of interest to a small class of people is now made available to all. As we have seen this has been a rather difficult encounter for geography. It sprawls between institutions, never quite able to be clipped into any one organisational shape. Academic geography has had a particularly rough ride: trying to turn geography into an ordinary academic specialism was never going to be easy. If academic historians had endured a similar experience to academic geographers – distracted by an elusive search for intellectual status and constant institutional fragmentation – they would probably have been afflicted by similar anxieties. Perhaps they would now be telling us that they were specialists in the study of 'temporal variation' and 'sociotemporal analysis'. But of course we know that history is not about windy abstractions: it is about the events and ideas of the past. Academic history has been more successful than academic geography in translating its pre-modern, singularly non-specialist sensibility ('the past' does, after all, cover everything) into the modern era. Academic geography may yet learn to live with itself in a similarly relaxed fashion.

Meanwhile, geography's institutions are evolving. New global organisations are emerging both to meet new challenges, such as environmental change, and new opportunities, such as the growing appetite for travel and travelogue. The dangers, freedoms and limitations of our technocentric, late modern era are shaping and reshaping the institutionalisation of the geographical imagination.

 What is Geography?

1   Geography is the world discipline.

2   Geography is rooted in the human need for survival; in the necessity of knowing and making sense of the resources and dangers of our human and physical environment. But it also seeks the bigger picture: geography helps us imagine that there is meaning and sense in the world. Geography allows us to see order in, and impose order on, what otherwise would be chaos.

3   Geography is both pre-modern and modern. It is a paradoxical and necessary combination. Geography's wide horizons and holistic sensibility are antithetical to an age of intellectual fragmentation and specialism. Yet a commitment to world knowledge is essential in a globalising era defined by environmental and political crises.

4   Geography has a wide subject matter and an equally wide constituency of contributors. Geography is both a popular and a professional activity. To hear geography's story we must listen to voices from across the world and from many different intellectual traditions.

5   In the modern era the geographical imagination has been structured into two basic tendencies, namely the pursuit of international and environmental knowledge. This division reflects modernity's global ambitions and its power to alienate people from nature (nature has been turned into something separate from people; a discrete arena for exploitation or wonder).

6   Industrial modernity has environmental and social consequences (such as environmental change, population growth, and pollution) that require geography's integrating, inclusive and global approach.

7   The modern era has produced rapid shifts in settlement activity. These have, in turn, provoked points of focus for the modern geographical imagination, notably urbanisation and mobility.

8   Geography has a number of distinctive practices. These include mapping, exploration, and connecting and combining knowledge about human and natural systems. Geography's outward disposition also encourages an engaged, involved outlook; a desire not merely to observe the world but to change it for the better.

9   Geography is an attempt to both understand and meet the world. In this sense, we may say that geographers are explorers. This ambition makes geography a distinctive contribution to increasingly bureaucratic and institutionalised systems of education. Its wide sweep, its long history and its curiosity about the world outside the window, make geography an awkward discipline for the classroom or lecture theatre. Geography's difficult institutional history within academia indicates how hard it is to mould a non-specialist activity into something that resembles conventional specialisms. However, the institutionalisation process has also been central to opening geography to wider audiences. It appears that geography must exist within and against its modern institutional forms.

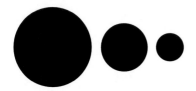

Notes

Introduction

1 Peter Bowler, *The Fontana History of the Environmental Sciences*, Fontana, London, 1992, p. 391.
2 'The Blue Marble', en.wikipedia.org/wiki/Blue_Marble, accessed 27/06/2006. See also Denis Cosgrove, *Apollo's Eye: A Cartographic Genealogy of the Earth in the Western Imagination*, Johns Hopkins University Press, Baltimore, 2001.
3 See Joseph May, *Kant's Concept of Geography and its Relation to Recent Geographical Thought*, University of Toronto Press, Toronto, 1970.
4 Edward Hallet Carr, *What is History?*, Penguin, London, 1987, p. 8.
5 Serge Latouche, *The Westernization of the World: Significance, Scope and Limits of the Drive Towards Global Uniformity*, Polity Press, Oxford, 1996.
6 See also Alastair Bonnett, 'Geography as the world discipline: connecting popular and academic geographical imaginations', *Area*, 35, 1, 2003 pp. 55–63.

Chapter 1

1 James Romm, *The Edges of the Earth in Ancient Thought: Geography, Exploration and Fiction*, Princeton, Princeton University Press, 1992.
2 Lucius Seneca, *Medea*, Oxford University Press, Oxford, 1973, p. 36. Tethys is sister and wife of Oceanus, personification of the sea.
3 Cf. David Stoddart, *On Geography*, Blackwell, Oxford, 1986. For Stoddart geography is a uniquely European creation: 'In method and in concept geography as we know it today is overwhelmingly a European discipline. It emerged as Europe encountered the rest of the world, and indeed itself, with the tools of the new objective science, and all other

geographical traditions are necessarily derivative and indeed imitative of it'. p. 39.

4   *The Holy Bible*, Psalms, 19, 1.

5   Ibid., Genesis, 1, 26.

6   See John Kirtland Wright, *The Geographical Lore of the Time of the Crusades: A Study in the History of Medieval Science and Tradition in Western Europe*, Dover Publications, New York, 1965.

7   Quoted in Wright, *The Geographical Lore*, 1965, p. 270.

8   John Kirtland Wright, *The Geographical Lore of the Time of the Crusades: A Study in the History of Medieval Science and Tradition in Western Europe*, Dover Publications, New York, 1965.

9   See David Livingstone, *The Geographical Tradition: Episodes in the History of a Contested Enterprise*, Blackwell, Oxford, 1992.

10   Quoted in Margarita Bowen, *Empiricism and Geographical Thought: From Francis Bacon to Alexander von Humbolt*, Cambridge University Press, Cambridge, 1981, p. 110.

11   Quoted in Bowen, *Empiricism and Geographical Thought*, p. 277.

12   Quoted in Livingstone, *The Geographical Tradition*, 1992, p. 86.

13   Rudyard Kipling, *Sixty Poems*, Hodder and Stoughton, London, 1939, p. 11.

14   Quoted in Bernard Porter, *Critics of Empire: British Radical Attitudes to Colonialism in Africa 1895–1914*, Macmillan, London, 1968, p. 50.

15   Benjamin Kidd, *Social Evolution*, Macmillan, London, 1894, p. 317.

16   Michael Mandelbaum, *The Ideas that Conquered the World: Peace, Democracy, and Free Markets in the Twenty-first Century*, Public Affairs, New York, 2002.

17   Victor Hanson, *Why the West Won: Carnage and Culture from Salamis to Vietnam*, Faber and Faber, London, 2001.

18   John Roberts, *The Triumph of the West*, BBC, London, 1985.

19   Ibid., p. 431

20   Francis Fukuyama, *The End of History and the Last Man*, Hamish Hamilton, London, 1992, p. 136.

21   Benjamin Barber, *Jihad vs McWorld: How Globalism and Tribalism are Reshaping the World*, Ballantine Books, New York, 1996, p. 4.

22   Quoted in Clarence Glacken, *Traces on the Rhodian Shore: Nature and Culture in Western Thought from Ancient Times to the End of the Eighteenth Century*, University of California Press, Berkeley, 1976, p. 277.

23   'East Asian History Sourcebook: *Chinese Accounts of Rome, Byzantium and the Middle East*, c., 91 BCE–1643 CE, http://depts.washington.edu/silkroad/texts/romchin1.html, accessed on 20/12/2006.

24   Georg Hegel, *The Philosophy of History*, Prometheus Books, Amherst, 1991, p. 103.

25   Edward Said, *Orientalism: Western Representations of the Orient*, Penguin, Harmondsworth, 1978.

26 Quoted in Edward Hallet Carr, *A History of Soviet Russia: Socialism in One Country 1924–1926: Volume* One, Macmillan, London, 1958, p. 144.

27 *The Holy Bible*, Ezekiel, 5, 5.

28 'East Asian History Sourcebook: *Chinese Accounts of Rome, Byzantium and the Middle East*, c. 91 BCE–1643 CE, http://depts.washington.edu/silkroad/texts/romchin1.html, accessed on 20/12/2006.

29 Pliny the Elder, *Natural History: A Selection*, Penguin, London, 1991, p. 75

30 Strabo, 'The *Geography* of Strabo', http://penelope.uchicago.edu/Thayer/E/Roman/Texts/Strabo/1B2*.html, accessed 20/12/2006.

31 The name 'Strabo' may have been a nickname, as it is the Latin term for someone with an eye deformity or squint.

32 Strabo, 'The *Geography* of Strabo', http://penelope.uchicago.edu/Thayer/E/Roman/Texts/Strabo/2A1*.html and 'The *Geography* of Strabo', http://penelope.uchicago.edu/Thayer/E/Roman/Texts/Strabo/4E*.html, accessed 20/12/2006.

33 Strabo, 'The *Geography* of Strabo', http:// penelope.uchicago.edu/Thayer/E/Roman/Texts/Strabo/1A*.html, accessed 20/12/2006.

34 Strabo, 'The *Geography* of Strabo', http://penelope.uchicago.edu/Thayer/E/Roman/Texts/Strabo/7C*.html, accessed 20/12/2006.

35 W. Yeats, 'Introduction', in R. Tagore, *Gitanjali (Song Offerings)*, Macmillan, London, 1913, p. xx.

36 Quoted in F. Rosemont, 'Surrealists on whiteness from 1925 to the present', *Race Traitor*, 9, 1988, p. 7.

37 Henry Morton Stanley, 'Central Africa and the Congo Basin; or, the importance of the scientific study of geography', *Journal of the Manchester Geographical Society*, 1, pp. 1–3, 6–25, 1885, p. 8.

38 Ibid., p. 14.

39 Peter Davis, *The Flora of Turkey and the East Aegean Islands: Volume One*, Edinburgh University Press, Edinburgh, 1984.

40 Richard Grimmett, *Birds of India: Pakistan, Nepal, Bangladesh, Bhutan, Sri Lanka and the Maldives*, Princeton University Press, Princeton, 1999.

41 Isaiah Bowman, *The New World: Problems in Political Geography: Fourth Edition,* World Book Company, Yonkers-on-Hudson, New York, 1928, p. 6.

42 Peter Kropotkin, 'What geography ought to be', in J. Agnew, D. Livingstone and A. Rogers (Eds), *Human Geography: An Essential Anthology*, Blackwell, Oxford, 1996, p. 141.

43 Michel de Montaigne, *The Complete Essays*, Penguin Books, London, 1993, p. 231.

44 Jean-Jacques Rousseau, *A Discourse on Inequality*, Penguin Books, London, 1984, p. 159.

45 Fukuzawa Yukichi, *An Outline of a Theory of Civilization*, Sophia University, Tokyo, 1973, p. 99.
46 See Jane Leonard, *Wei Yuan and China's Rediscovery of the Maritime World*, Harvard, University Press, 1984; Frank Dikotter, *The Discourse of Race in Modern China*, Stanford University Press, Stanford, 1992.
47 Quoted in Dikotter, *The Discourse of Race in Modern China*, 1992, p. 160.
48 Ibid.
49 Gerard Delanty (Ed.), *Europe and Asia Beyond East and West*, Routledge, London, 2006.

Chapter 2

1 Clarence Glacken, *Traces on the Rhodian Shore: Nature and Culture in Western Thought from Ancient Times to the End of the Eighteenth Century*, University of California Press, Berkeley, 1976.
2 Ibid., p. 3.
3 For details of Eratosthenes's calculation see The Eratosthenes Project, at http://www.phys-astro.sonoma.edu/observatory/eratosthenes/, accessed 21.12.2006.
4 Quoted in Glacken, *Traces on the Rhodian Shore*, 1976, p. 87.
5 Strabo, 'The *Geography* of Strabo', http://penelope.uchicago.edu/Thayer/E/Roman/Texts/Strabo/2C*.html, accessed 20/12/2006.
6 Alfred Hettner, *Die Geographie, ihre Geschichte, ihr Wesen und ihre Methoden*, Ferinand Hirt, Breslau, 1927. Lucien Febvre, *A Geographical Introduction to History*, Knopf, London, 1925.
7 Febvre, *A Geographical Introduction*, 1925.
8 Jared Diamond, *Guns, Germs and Steel: A Short History of Everybody for the Last 13,000 Years*, Vintage, London, 1998. Alfred Crosby, *Ecological Imperialism: The Biological Expansion of Europe, 900–1900*, Cambridge University Press, Cambridge, 1986.
9 Jared Diamond, *Collapse: How Societies Choose to Fail or Succeed*, Allen Lane, London, 2005.
10 Jared Diamond, *Guns, Germs and Steel*, 2005, p. 32.
11 Quoted in Livingstone, *The Geographical Tradition*, 1992, p. 136.
12 Jean Brunhes, *Human Geography*, Rand McNally, Chicago, 1920, p. 15.
13 Quoted in George Tatham, 'Environmentalism and possibilism', in Griffith Taylor (Ed.), *Geography in the Twentieth Century: A Study of Growth, Fields, Techniques, Aims and Trends* (Third Edition), Methuen, London, 1957, p. 152.
14 Derek Gregory, *Ideology, Science and Human Geography*, Hutchinson, London, 1978, pp. 170–1.

15 William Wordsworth, 'Lines Written a Few miles above Tintern Abbey, on Revisiting the Banks of the Wye During a Tour, 13 July 1798', in Duncan Wu, *Romanticism: An Anthology*, Blackwell, Oxford, 1994, p. 242.

16 George Parkins Marsh, *Man and Nature*, Harvard University Press, Cambridge, 1965, p. 186.

17 Aldo Leopold, *Sand County Almanac: with other Essays on Conservation from Round River*, Oxford University Press, New York, 1966, p. x.

18 Brian Hayden, 'Subsistence and ecological adaptations of modern hunter-gatherers', in R. Harding and G. Teleki (Eds), *Omnivorous Primates*, Columbia University Press, New York, 1981.

19 Tu Wei-ming, 'The ecological turn in new Confucian humanism: implications for China and the world', *Daedalus* 130, 4, 2001, pp. 243–264. See also P. Harris, '"Getting rich is glorious": environmental values in the People's Republic of China', *Environmental Values* 13, 2, 2004, pp. 145–165.

20 Donella Meadows, Dennis Meadows, Jorgen Randers and William Behrens, *Limits to Growth: A Report For The Club Of Rome's Project On The Predicament Of Mankind*, A Potomac Associates Book, New York, 1972.

21 Ernst Schumacher, *Small Is Beautiful. A Study Of Economics As If People Mattered*, Blond and Briggs, London, 1973.

22 Barry Commoner, *The Closing Circle: Nature, Man and Technology*, Knopf, New York, 1971, p. 1.

23 Barry Commoner, 'Fundamental causes of the environmental crisis', in Roderick Nash (Ed.), *American Environmentalism: Readings in Conservation History, Third Edition,* McGraw-Hill, New York, 1990, p. 206.

24 Stephen Cotgrove and Andrew Duff, 'Environmentalism, middle-class radicalism and politics', *Sociological Review*, 28, 2, 1980, pp. 333–51.

25 Rachel Carson, *Silent Spring*, Houghton Mifflin, Boston, 1962, p. 12.

26 Dave Foreman, quoted in Bill McKibben, *The End of Nature: Humanity, Climate Change and the Natural World*, Bloomsbury, London, 2003, p. 194.

27 Ian Sample, 'Forests are poised to make a comeback, study shows', *The Guardian*, 14 November 2006.

28 Spencer Weart, *The Discovery of Global Warming*, Harvard University Press, Cambridge, 2003.

29 Svante Arrhenius, *Worlds in the Making,* Harper & Brothers, New York, 1908, p. 63.

30 Nils Ekholm, 'On the variations of the climate of the geological and historical past and their causes', *Quarterly J. Royal Meteorological Society* 27, 1901, p. 61.

31   Willi Dansgaard, Sigfus Johnsen, Henrik Clausen, Darthe Dahl-Jensen, Niels Gundestrup, Claus Hammer, Christine Huindberg, Jorgen Steffensen, Arny Sveinbjörnsdottir, Jean Jouzel and Gerard Bond 'Evidence for general instability of climate from a 250–kyr ice-core record', *Nature* 364, 1993, pp. 218–220. See also Weart, *The Discovery of Global Warming,* 2003.

32   Bjon Lomborg, *The Sceptical Environmentalist*, Cambridge University Press, Cambridge, 2001.

33   International Panel on Climate Change, *Climate Change 2007: Impacts, Adaptation and Vulnerability: Working Group II Contribution to the Intergovernmental Panel on Climate Change Fourth Assessment Report: Summary for Policymakers,* IPCC, Geneva, 2007, p. 2.

34   HM Treasury, *The Stern Review on the Economics of Climate Change*, HM Treasury, London, 2006.

35   Urs Siegenthaler, Thomas Stocker, Eric Monnin, Dieter Lüthi, Jakob Schwander, Bernhard Stauffer, Dominique Raynaud, Jean-Marc Barnda, Hubertus Fischer, Valérie Masson-Delmotte and Jean Jouzel, 'Stable carbon cycle-climate relationship during the late Pleistocene', *Science*, 310, 2005, pp. 1313–1317.

36   Tim Barnett, David Pierce, Krishna AchutaRao, Peter Gleckler, Benjamin Santer, Jonathan Gregory and Warren Washington 'Penetration of human-induced warming into the world's oceans', *Science*, 309, 2005, pp. 284–287.

37   HM Treasury, *The Stern Review.*

38   Intergovernmental Panel on Climate Change, *Climate Change 2001: Summary for Policy Makers,* Cambridge University Press, Cambridge, 2001.

39   Katey Walter, Sergey Zimou, Jeff Chanton, Dave Verbyla and Terry Chapin, 'Methane bubbling from Siberian thaw lakes as a positive feedback to climate warming', *Nature*, 443, 2006, pp. 71–75.

40   HM Treasury, *Stern Review,* p. 12.

41   HM Treasury, *Stern Review,* p. 56.

42   Norman Myers, 'Environmental Refugees: An Emergent Security Issue', paper delivered at 13th Economic Forum, Prague, 23–27 May, 2005.

43   Bhimanto Suwastoyo, 'Haze from Indonesia fires chokes region, spreads across pacific', http://www. terradaily.com/reports/Haze_From_ Indonesia_Fires_Chokes_Region_Spreads_Across_Pacific_999.html, accessed 21.12.2006.

44   Comparative Risk Assessment Collaborating Group, 'Selected major risk factors and global and regional burden of disease', *Lancet*, 360, 2002, pp. 1347–1360.

45   Andrew Goudie and Heather Viles, *The Earth Transformed: An Introduction to Human Impacts on the Environment*, Blackwell, Oxford, 1997.

46 Daniel Nepstad, Paulo Moutinho and Britaldo Soares-Filho, *The Amazon in a Changing Climate: Large-Scale Reductions of Carbon Emissions from Deforestation and Forest Impoverishment*, http://www. whrc.org/resources/published_literature/pdf/ Amazon-and-Climate-2006.pdf, accessed 8.4.2007.

47 Graham Harvey, *We Want Real Food*, Constable, London, 2006 p. 171.

48 Vaclav Smil, *Enriching the Earth*, MIT Press, Cambridge, 2001.

49 Paul Ehrlich and John Holdren, 'Impact of population growth', *Science*, 171, 1971, pp. 1212–1217. John Holdren and Paul Ehrlich, 'Human population and the global environment, *American Scientist*, 62, 3, 1974, pp. 282–292.

50 Georgine Mace, Hillary Masundire and Jonathan Baillie, 'Biodiversity', in Rashid Hassan, Robert Scholes and Neville Ash (Eds) *Ecosystems and Human Well-Being: Volume 1*, Island Press, Washington, 2005, p. 109.

51 Ibid., p. 79.

52 UN News Center, 'World population to reach 9.1 billion in 2050, UN projects', http://www.un.org/apps/news/story.asp?NewsID=13451&Cr= population&Cr1, accessed 22.12.2006.

53 Thomas Malthus, *An Essay on the Principle of Population*, Harmondsworth, Pelican, 1976, p. 71.

54 Academy of Sciences, *Population Summit of the World's Scientific Academies*, National Academy Press, Washington, 1993.

55 William Turner, Robert Hanham and Anthony Portararo, 'Population pressure and agricultural intensity', *Annals of the Association of American Geographers*, 67, 1977, pp. 386–97.

56 Ester Boserup, *Population and Technology*, Blackwell, Oxford, 1980.

57 Quoted in Clarence Glacken, *Traces on the Rhodian Shore*, 1976, p. 65.

58 Aristotle, *Meteorologica,* Harvard University Press, Cambridge, 1952. See also Pliny the Elder's discussion of the 'eight main winds' in *Natural History: A Selection*, Penguin Books, London, 1991.

59 Svante Arrhenius, 'On the influence of carbonic acid in the air upon the temperature of the round', *London, Edinburgh, and Dublin Philosophical Magazine and Journal of Science*, 41, 1896, pp. 237–275.

60 Milutin Milankovitch, *Theorie Mathematique des Phenomenes Thermiques produits par la Radiation Solaire*, Gauthier-Villars, Paris, 1920. *Milan Milankovitch, Mathematische Klimalehre und Astronomische Theorie der Klimaschwankungen, Handbuch der Klimalogie Band 1*, Teil A Borntrager, Berlin, 1930.

61 Edward Lorenz, *Nonlinearity, Weather Prediction, and Climate Deduction*, Massachusetts Institute of Technology, Dept of Meteorology, Cambridge, 1966. Edward Lorenz, *The Essence of Chaos*, University of Washington Press, Seattle, 1996.

62 James Hutton, 'THEORY of the EARTH; or an INVESTIGATION of the Laws observable in the Composition, Dissolution, and Restoration of Land

upon the Globe', http://www.mala.bc.ca/~Johnstoi/essays/Hutton.htm, accessed 22.12.2006.

63 Ibid.
64 Ibid.
65 Ibid.
66 Alfred Wegener, *The Origin of Continents and Oceans*, Dover Publications, New York, 1966.
67 Quoted in 'Alfred Wegener vs just about everybody else: how the continents formed (1912–1960's) http://courses.science.fau.edu/ ~rjordan/ phy1931/WEGENER/wegener. pdf, accessed 22.12.2006.
68 Halford Mackinder, 'On the scope and methods of geography', in John. Agnew, David Livingstone and Alisdair Rogers (Eds), *Human Geography: An Essential Anthology*, Blackwell, Oxford, 1996, p. 170.
69 Stephen Schneider and Penelope Boston (Eds), *Scientists on Gaia*, MIT Press, Cambridge, 1991, p. x.

Chapter 3

1 William Morris, *Political Writings: Contributions to Justice and Commonweal 1883–1890*, Thoemmes Press, Bristol, 1994, p. 25.
2 Department of Economic and Social Affairs, 'Population Division World Urbanization Prospects: The 2005 Revision', http://www. un.org/esa/population/publications/WUP2005/2005wup.htm,accessed 22.12.2006.
3 Earth Trends, 'Population, Health and Human Well-being – Urban and Rural Areas: Urban population as a percent of total population', http:// earthtrends.wri.org/text/population-health/variable-448.html, accessed 22.12.2006.
4 Ibid.
5 Ibid.
6 *People's Daily Online*, 'China encourages mass urban migration', http:// english.people.com.cn/200311/28/eng20031128_129252.shtml, accessed 22.12.2006.
7 Quoted in *People's Daily Online*, 'China encourages mass urban migration', http://english.people.com.cn/200311/28/eng20031128_ 129252.shtml, accessed 22.12.2006.
8 Lewis Mumford, *The City in History*, Harmondsworth, Penguin, 1966, p. 224.
9 Quoted in James Donald, 'This, here, now: imagining the modern city', in Sallie Westwood and John Williams (Eds), *Imagining Cities: Scripts, Signs, Memories*, Routledge, London, 1997, p. 195. Corbusier's stark

geometric plans for high rise cities were not the only solutions on offer. Although the semi-pastoral vision behind Ebernezer Howard's 'Garden Cities' was very different, it too was structured around a desire to achieve symmetry and the spatial division of urban functions. See Ebenezer Howard, *Garden Cities of To-morrow*, Swan Sonnenschein, London, 1902.

10 Arthur Nelson, 'By 2030 the US will have re-built almost half its built environment', http://www.citymayors.com/development/built_environment_usa.html, accessed 22.12.2006.

11 Ernest Burgess, 'The growth of the city: an introduction to a research project', in Robert Park, Ernest Burgess and R. McKenzie (Eds), *The City*, University of Chicago Press, Chicago, 1925.

12 John Rex, 'The sociology of a zone of transition', in Ray Pahl (Ed.) *Readings in Urban Sociology*, Pergamon, London, 1968, p. 214.

13 David Harvey, 'The urban process under capitalism: a framework for analysis', *International Journal of Urban and Regional Research*, 2, pp. 101–31, 1978, p. 120.

14 David Harvey, 'Geography policies, financial institutions and neighbourhood change in United States cities, in M. Harloe (Ed.) *Captive Cities*, John Wiley, London, 1977, p. 124.

15 David Harvey, *The Condition of Postmodernity. An Enquiry into the Origins of Cultural Change*, Oxford, Blackwell, 1990. This book was itself the subject of criticism from feminist writers, see Meahgan Morris, 'The Man in the Mirror: David Harvey's "Condition" of Postmodernity' *Theory, Culture Society*, 9, 1992, pp. 253–279. See also Noel Castree (Ed.) *David Harvey: A Critical Reader*, Blackwell, Oxford, 2006.

16 Michel de Montaigne, *The Complete Essays*, 1993, p. 235.

17 Ibid, pp. 240–1.

18 Ferdinand Tönnies, *Community and Society*, Harper and Row, New York, 1963. Georg Simmel, *Simmel on Culture: Selected Writings*, Sage, London, 1997.

19 Quoted in Peter Saunders, *Social Theory and the Urban Question*, Hutchinson, London, 1981, p. 89.

20 Richard Sennett, *Flesh and Stone: The Body and the City in Western Civilization*, Faber and Faber, London, 1994, pp. 25–26.

21 Guy Debord, *Panegyric*, Verso, London, 1991, pp. 44–45.

22 Guy Debord, *Society of the Spectacle*, Black and Red, Detroit, 1983, thesis 177.

23 Earth Trends, 'Population, health and human well-being – urban and rural areas: urban population as a percent of total population', http://earthtrends.wri.org/text/population-health/variable-448.html, accessed 22.12.2006.

24 Stanley Brunn and Jack Williams, *Cities of the World: World Regional Urban Development*, Harper and Row, New York, 1983, p. 36.

25   Mark Burkholder and Lyman Johnson, *Colonial Latin America*, Oxford University Press, New York, 1994, p. 175.

26   Jonathan Spencer, 'Occidentalism in the East: the uses of the West in the politics and anthropology of South Asia', in J. Carrier (Ed.) *Occidentalism: Images of the West*, Oxford University Press, Oxford, 1995.

27   Quoted in Ben Kierman, *The Pol Pot Regime: Race, Power, and Genocide in Cambodia under the Khmer Rouge, 1975–79*, New Haven, Yale University Press, 1996, p. 57.

28   Rabindranath Tagore, *Creative Unity*, Macmillan, London, 1922, p. 144.

29   Quoted in Stephen Hay, *Asian Ideas of East and West: Tagore and his Critics in Japan, China, and India*, Harvard University Press, Cambridge, 1970, p. 180.

30   See, for example, Maryam Jameelah, *Western Civilization Condemned by Itself: A Comprehensive Study of Moral Retrogression and its Consequences, Volume 1*, Mohammad Yusaf Khan and Sons, Lahore, 1979.

31   Ernesto Pernia (Ed.) *Urban Poverty in Asia: A Survey of Critical Issues*, OUP China, Hong Kong, 1999. See also Tim Bunnell, Lisa Drummond and Ho Kong Chong (Eds), *Critical Perspectives on Cities in Southeast Asia*, Times Academic Press, Singapore, 2002.

32   Lewis Mumford, *The City in History*, 1966, p. 616.

33   Jean Gottmann, *Megalopolis: The Urbanized North-Eastern Seaboard of the United States*, MIT Press, Cambridge, 1961.

34   Arif Dirlik, 'Place-based imagination: globalism and the politics of place', in Roxann Prazniak and Arif Dirlik (Eds), *Places and Politics in an Age of Globalization*, Lanham, Rowman and Littlefield, 2001, p. 42.

35   Edward Relph, *Place and Placelessness*, Pion, London, 1976; Marc Auge, *Non-Places: Introduction to an Anthropology of Supermodernity*, Verso, London, 1995; James Kunstler, *The Geography of Nowhere*, Simon and Schuster, New York, 1993. See also Jane Jacobs, *The Death and Life of Great American Cities*, Random House, New York, 1961.

36   Marc Auge, *Non-Places*, 1995, p. 110.

37   On these new class dynamics see Paul Cloke and Jo Little (Eds), *Contested Countryside Cultures*, Routledge, London. 1997; Jon Murdoch et al., *The Differentiated Countryside*, Routledge, London, 2003.

38   Loretta Lees, 'Urban geography: The "death" of the city?', in Alisdair Rogers and Heather Viles (Eds), *The Student's Companion to Geography: Second Edition*, Blackwell, London, 2003, p.127. Lees is summarising the views of Michael Dear, *The Postmodern Urban Condition*, Blackwell, Oxford, 2000.

39   James Heartfield, 'Londonostalgia', reprinted from *Blueprint*, September 2004, http://www.design4design.com/artucles/artcles_story. asp?STORYID=5765, accessed 05/04/2005.

40　Karl Marx, *The Revolutions of 1848*, Penguin, Harmondsworth, 1973, p. 71.

41　Harvey is critical 'of all those manifestations of place-bound nostalgias that infect our images of the country and the city, of region, milieu, and locality'. David Harvey, *The Condition of Postmodernity*, Blackwell, Oxford, 1989, p. 218.

42　Doreen Massey, 'A global sense of place', in Trevor Barns and Derek Gregory (Eds), *Reading Human Geography*, Arnold, London, 1997, p. xx. Massey develops her vision of a 'politics of place beyond place' in *World City*, Cambridge, Polity Press, 2007, p. 188. See also Tim Cresswell, *Place: A Short Introduction*, Blackwell, London, 2004; Linda McDowell (Ed.), *Undoing Place? A Geographical Reader*, Arnold, London, 1997.

43　Ibid. p. xx.

44　William Mitchell, *City of Bits: Space, Place and the Infobahn*, MIT Press, Cambridge, 1995, p. 161, p. 166.

45　Mark Gottdiener and Leslie Budd, *Key Concepts in Urban Studies*, London, Sage, 2005, p. 15.

46　Louis Wirth, 'Urbanism as a way of life', *American Journal of Sociology*, 44, pp. 1–24, 1938, p. 12.

47　Goran Rystad, 'Immigration history and the future of international migration', *International Migration Review*, 26, 4, pp. 1168–1199, 1992, p. 1170.

48　National Latino Statistics, http://www.justicebychoice.org/National%20Latino%20Statistics.pdf, accessed 26.12.2006.

49　Lindsay Lowell, 'Immigrants and labor force trends: the future, past, and present', http:// www.migrationpolicy.org/ITFIAF/TF17_ Lowell.pdf, accessed 26.12.2006.

50　Paul E. Lovejoy, *Transformations in Slavery*, Cambridge University Press, Cambridge, 2000.

51　Pavel Polyan, *Не по своей воле...История и география принудительных миграций в ССР,* (translation: 'Not by their own will... a history and geography of forced migrations in the USSR'), *ОГИ Мемориал ,* Moscow, 2001. See also Population transfer in the Soviet Union, http://en. wikipedia.org/wiki/Population_transfer_in_the_Soviet_Union, accessed 26.12.2006. Terry Martin, 'The origins of Soviet ethnic cleansing', *Journal of Modern History*, 70, 1998, pp. 813–861.

52　George Ravenstein, 'The laws of migration', *Journal of the Royal Statistical Society*, 52, 2, 1889, pp. 241–305. See also Waldo Tobler, 'Migration: Ravenstein, Thorntwaite, and beyond', *Urban Geography*, 16, 4, 1995, pp. 327–343.

53　Tomas Hammer, *Democracy and the Nation State: Aliens, Denizens and Citizens in a World of International Migration*, Avebury, Aldershot, 1990.

54 Adrian Bailey, *Making Population Geography*, Hodder Arnold, London, 2005, p. 125. Bailey is drawing on the work of Doreen Mattingly, 'The home and the world: domestic service and international networks of caring labor', *Annals of the Association of American Geographers*, 91, pp. 370–86, 2001.

55 Warwick Murray, *Geographies of Globalization*, Routledge, London, 2006.

56 *Ibid.*

57 Folker Frobel, Jurgen Heinrichs and Otto Krege, *The New International Division of Labour*, Cambridge University Press, Cambridge, 1980.

58 Robert Gwynne, Thomas Klak and Denis Shaw, *Alternative Capitalisms: Geographies of Emerging Regions*, Arnold, London, 2003, p. 168.

59 John Bryson and Nick Henry, 'The global production system: from Fordism to post-Fordism', in Peter Daniels, Michael Bradshaw, Denis Shaw and James Sidaway (Eds), *Human Geography: Issues for the 21st Century*, Prentice Hall, London, 2001, p. 369. See also Saskia Sassen, *The Global City: London, New York, Tokyo*, Princeton University Press, Princeton, 2001.

60 Corporate History, http://www.gm.com/company/corp_info/history/gmhis1920.html, accessed 26.12.2006.

61 Olle Hagman, 'Morning queues and parking problems: on the broken promises of the automobile', *Mobilities*, 1, 1, pp. 63–74, p. 67.

62 Luis de la Fuente Layos, *Statistics in Focus: Transport: Short Distance Passenger Mobility in Europe*, Eurostat, Luxembourg, 2005.

63 Scottish Executive, 'Statistical Bulletin Transport Series: Trn/2005/3: Travel by Scottish residents: some National Travel Survey results for 2002/2003 and earlier years', http://www.scotland.gov.uk/Publications/2005/04/1894658/46593, accessed 28.12.2006.

64 Brian Handwerk, 'China's Car Boom Tests Safety, Pollution Practices', http://geography.about.com/gi/dynamic/offsite.htm?site=http://news.nationalgeographic.com/news/2004/06/0628%5F040628%5Fchinacars.htm, accessed 27.12.2006.

65 'Delphi expects $1 billion China sales', http://www.china-defense.com/forum/ showthread.php?t=8620&page=6, accessed 27.12.2006.

66 The Chinese Outbound Tourism Market, http://www.world-tourism.org/newsroom/Releases/2006/november/chineseoutbound.htm, accessed 27.12.2006.

67 Dan MacCannell, *Empty Meeting Grounds: The Tourist Papers*, Routledge, London, 1992.

68 Rabindranath Tagore, *Creative Unity,* Macmillan, London, 1922, p. 95.

69 Quoted in Geoffrey Martin and Preston James, *All Possible Worlds: A History of Geographical Ideas*, John Wiley, London, 1993, p. 440.

70 Quoted in Goran Rystad, 'Immigration history', 1992, p. 1172.

Chapter 4

1 Halford Mackinder, 'On the scope and methods of geography', 1996, p. 156.

2 Stan Stevens, 'Fieldwork as commitment', *The Geographical Review*, 91, 2001, pp. 66–73, p. 66.

3 Bronislaw Malinowski, *Argonauts of the Western Pacific*, 1922, George Routledge and Sons, London.

4 Heidi Nast, 'Opening remarks on "women in the field"', *Professional Geographer*, 46, 1994, pp. 54–66, p. 56.

5 Gillian Rose, *Feminism and Geography*, Polity Press, Cambridge, 1993, p. 109.

6 Cited in Stoddard, *On Geography*, p. 47.

7 Teresa Ploszajska, *Geographical Education, Empire and Citizenship: Geographical Teaching and Learning in English Schools, 1870–1944*, Historical Geography Research Group, London, 1999, p. 270.

8 Livingstone, *The Geographical Tradition*, 1992, p. 129.

9 Montesquieu, *Persian Letters*, Penguin, London, 1973, p. 239.

10 Malinowski, *Argonauts of the Western Pacific*, 1922, pp. 21–22.

11 James Clifford, *The Predicament of Culture: Twentieth-Century Ethnography, Literature, and Art*, Harvard University Press, Cambridge, 1988; Clifford Geertz, *Works and Lives: The Anthropologist as Author*, Stanford University Press, Stanford, 1988.

12 See, for example, http://www.travelblogs.com/.

13 See, for example, *Transgressions: A Journal of Urban Exploration* and the 'urban explorers' networks at http://www.urbanexplorers.net/ and http://www.sub-urban.com/fleettwo.htm.

14 Iain Sinclair, *London Orbital*, Penguin, London, 2003. See also Iain Sinclair, *London: City of Disappearances*, Hamish Hamilton, London, 2006.

15 Steve Watkins and Clare Jones, *Unforgettable Journeys to Take Before You Die*, BBC Books, London, 2000. Patricia Schultz, *1000 Places to See Before you Die*, Workman Publishing, New York, 2003. Steve Davey, *Unforgettable Places to See Before you Die*, BBC Books, London, 2004.

16 Carl Ritter, *Comparative Geography*, William Blackwood and Sons, Edinburgh, 1865, p. 16.

17 Halford Mackinder, 'On the scope and methods of geography', 1996, p. 159.

18 See, for example, Rob Kitchin and Nicholas Tate, *Conducting Research into Human Geography*, Prentice Hall, Harlow, 2000 and Guy Robinson, *Techniques and Methods in Human Geography*, Wiley, London, 1998. Cf. John Matthews and David Herbert, *Unifying Geography: Common Heritage, Shared Future*, Routledge, London, 2004.

19 Hugh Lamprey, 'Pastoralism yesterday and today: the over-grazing problem', in Francois Bourliere (Ed.), *Tropical Savannas: Ecosystems of the World, Volume 13*, Elsevier, London, 1981.

20 James Lovelock, *Gaia: A New Look at Life on Earth*, Oxford Paperbacks, Oxford, 2000.

21 Richard Hartshorne, *The Nature of Geography: A Critical Survey of Current Thought in the Light of the Past*, Association of American Geographers, Lancaster, 1961, p. xii.

22 See, for example, Vernon Meentemeyer, 'Geographical perspectives of space, time, and scale', *Landscape Ecology*, 3, 3–4, 1989, pp. 163–173. Yehua Dennis Wei, 'Multiscale and multimechanisms of regional inequality in China: implications for regional policy', *Journal of Contemporary China*, 11, 30, 2002, pp. 109–124.

23 Olaf Bastian, 'Landscape ecology – towards a unified discipline', *Landscape Ecology*, 16, 2002, pp. 757–766, p. 764.

24 Elazar Barkan, *The Retreat from Scientific Racism*, Cambridge University Press, Cambridge, 1992.

25 Peter Haggett, *The Geographer's Art*, Blackwell, Oxford, 1995.

26 Richard Hartshorne, *The Nature of Geography*, 1961, p. xi.

27 Peter Gould and Rodney White, *Mental Maps*, Routledge, London, 1985.

28 Jonathan Swift, *On Poetry, a Rapsody*, n.p. , Dublin and London, 1733.

29 Jeremy Black, *Visions of the World: A History of Maps*, Mitchell Beazley, London, 2003, p. 27.

30 Quoted in Harm de Blij, *Why Geography Matters: Three Challenges Facing America: Climate Change, the Rise of China, and Global Terrorism*, Oxford University Press, Oxford, 2005, p. 48.

31 Quoted in Noel Castree, 'Whose geography? Education as politics', in Noel Castree, Alisdair Rogers and Douglas Sherman (Eds), *Questioning Geography*, Blackwell, Oxford, 2005, p. 297.

32 Quoted in Livingstone, *The Geographical Tradition*, 1992, p. 104.

33 Guy Bessette, *Involving the Community: A Guide to Participatory Development Communication*, International Development Research Centre, Ottawa, 2004.

34 Trevor Wickham, 'Farmers ain't no fools: exploring the role of participatory rural appraisal to access indigenous knowledge and enhance sustainable development research and planning. A case study of Dusun Pausan, Bali, Indonesia', Master's Thesis, Faculty of Environmental Studies, University of Waterloo, 1993.

35 Louise Grenier, *Working with Indigenous Knowledge: A Guide for Researchers*, International Development Research Centre, Ottawa, 1998.

36 See, for example, William Bunge, *Fitzgerald: Geography of a Revolution*, Schenkman Publishing, Cambridge, 1971.

37 Edward Abbey, 'Foreward!', in Dave Foreman (Ed.), *Ecodefense: A Field Guide to Monkeywrenching*, Ned Ludd, Tucson, 1987, p. 7.

38 News from Nowhere (Ed.) *We Are Everywhere: The Irresistible Rise of Global Anti-capitalism*, Verso, London, 2003; Paul Kingsnorth *One No,*

Many Yeses: A Journey to the Heart of the Global Resistance Movement,
Free Press, London, 2004.

39 Karen Malone and Paul Tranter, 'School grounds as sites of learning:
making the most of environmental opportunities', *Environmental
Education Research*, 9, 3, 2003, pp. 283–303, p. 284. See also Gary
Habhan and Stephen Trimble, *The Geography of Childhood: Why
Children Need Wild Places*, Beacon Press, Boston 1994.

Chapter 5

1 Clarence Glacken, *Traces on the Rhodian Shore*, 1976, p. xiii.
2 Quoted in Peter Haggett, *The Geographer's Art*, 1995, p. 129.
3 *Journal of the Royal Geographical Society of London*, 1, 1831, p. vii.
4 Quoted in Om Kejariwal, *The Asiatic Society of Bengal and the Discovery
of India's Past*, Oxford University Press, New Delhi, 1999, p. 35.
5 It is interesting to observe how the regions that attract research funding and interest at any one time are usually those which are also the
objects of political and economic attention. Noting the development of
state funding for Soviet, Russian and Japanese studies in universities
across the USA after the Second World War, Miyoshi and Harootunian
explain that it was designed 'to meet the necessity of gathering and
providing information about the enemy'. See Masao Miyoshi and Harry
Harootunian, 'Introduction: the "afterlife" of Area Studies', in Masao
Miyoshi and Harry Harootunian (Eds), *Learning Places: The Afterlives
of Area Studies*, Duke University Press, Durham, 2002, p. 2.
6 John Nietz, *Old Textbooks: Spelling, Grammar, Reading, Arithmetic,
Geography, American History, Civil Government, Physiology,
Penmanship, Art, Music, as Taught in the Common Schools from Colonial
Days to 1900*, University of Pittsburgh Press, Pittsburgh, 1961, p. 196.
7 John Richard Green, 'Introduction', in J.R. Green and Alice Green,
A Short Geography of the British Isles, Macmillan, London, 1879,
pp. vii–viii.
8 Quoted in Stoddart, *On Geography*, 1986, p. 183.
9 Quoted in Stoddart, *On Geography*, 1986, p. 83.
10 James Conant quoted in Neil Smith, '"Academic wars over the field of
geography": the elimination of geography at Harvard, 1947–1952', *Annals
of the Association of American Geographers*, 77, 1982, pp. 155–72, p. 159.
11 See, for example, Livingstone, *The Geographical Tradition,* 1992.
12 Quoted in Livingstone, *The Geographical Tradition* , 1992, p. 204.
13 Nevin Fenneman, 'The circumference of geography', *Annals of the
Association of American Geographers*, 9, 1919, pp. 3–11.

14   Isaiah Bowman, *Geography in Relation to the Social Sciences*, Charles Scribner's Sons, New York, 1934, p. 146.

15   Paul Vidal de la Blache, *Principes de geographie humaine*, Colin, Paris, 1922.

16   Peter Haggett, *The Geographer's Art*, 1995, p. 79.

17   George Kimble, 'The inadequacy of the regional concept', in Laurence Dudley Stamp and Sidney Williams Wooldridge (Eds), *London Essays in Geography*, Longmans, Green, London, 1951; John Paterson, 'Writing regional geography: problems and progress in the Anglo-American realm', *Progress in Human Geography*, 6, 1974, pp. 1–16.

18   See, for example, Vernon Meentemeyer, 'Geographical perspectives of space, time, and scale', 1989. Yehua Wei, 'Multiscale and multi mechanisms of regional inequality in China', 2002.

19   Richard Hartshorne, *Perspectives on the Nature of Geography*, Rand McNally, Chicago, 1959, p. 21.

20   Peter Haggett, *The Geographer's Art*, 1995, p. 8.

21   Hence the need for attempts at 'reunification', for example, John Matthews and David Herbert, *Unifying Geography: Common Heritage, Shared Future*, Routledge, London, 2004.

22   Richard Hartshorne, *The Nature of Geography*, 1961, p. 460.

23   David Harvey, *The Limits to Capital*, Basil Blackwell, Oxford, 1982.

24   Michael Dear and Steven Flusty, 'Introduction', in M. Dear and S. Flusty (Eds), *The Spaces of Postmodernity*, Blackwell, Oxford, 2002, p. 2.

25   Doreen Massey, *For Space*, Sage, London, 2005, p. 91.

26   Ibid., p. 189.

27   See, for example, Harry Harootunian, *Overcome by Modernity: History, Culture and Community in Interwar Japan*, Princeton University Press, Princeton, 2000. Shmuel Eisenstadt, 'Multiple modernities', *Daedalus*, 129, 1, 2000, pp. 1–29. Alastair Bonnett, 'Occidentalism and plural modernities: or how Fukuzawa and Tagore invented the West', *Environment and Planning D: Society and Space*, 23, 2005, pp. 505–525.

28   Nicky Gregson, 'Discipline games, disciplinary games and the need for a post-disciplinary pratice: responses to Nigel Thrift's "The future of geography"', *Geoforum*, 34, 2003, pp. 5–7.

29   Georges Benko and Ulf Strohmayer, 'Conclusion, or an introduction to human geography in the 21st century', in Georges Benko and Ulf Strohmayer (Eds), *Human Geography: A History for the 21st Century*, Arnold, London, 2004, p. 139.

30   Robert Stafford quoted in Livingstone, *The Geographical Tradition*, 1992, p. 169.

31   Quoted in Kirkpatrick Sale, 'Schism in environmentalism'; *Roderick Nash* (Ed.), *American Environmentalism: Readings in Conservation History; Third Edition*, McGraw-Hill, New York, 1990, p. 285.

32 See Shamima Ahmed and David Potter, *NGOs in International Politics*, Kumarian Press, Bloomfield, CT, 2006; Julie Fisher, *Nongovernments: NGOs and the Political Redevelopment of the Third World*, Kumarian Press, Bloomfield, CT, 1998.

33 See Paul Smith, *The History of Tourism: Thomas Cook and the Orgins of Leisure Travel*, Routledge, London, 1998; Harmut Berghoff et al. (Eds), *Making of Modern Tourism: The Cultural History of the British Experience, 1600–2000*, Palgrave, London, 2002.

34 See Donald Read, *The Power of News: The History of Reuters: Second Edition*, Oxford University Press, Oxford, 1999; Stanley Baran and Roger Wallis, *The Known World of Broadcast News: International News and the Electronic Media*, Routledge, London, 1990.

35 Joseph Conrad, *Last Essays*, J. M. Dent, London, 1926, p. 17.

 Bibliography

Abbey, E. (1987) 'Foreward!', in Dave Foreman (Ed.) *Ecodefense: A Field Guide to Monkeywrenching*, Tucson: Ned Ludd.

Academy of Sciences (1993) *Population Summit of the World's Scientific Academies*, Washington, DC: National Academy Press.

Ahmed, S. and Potter, D. (2006) *NGOs in International Politics*, Bloomfield, CT: Kumarian Press.

Aristotle (1952) *Meteorologica,* Cambridge, MA: Harvard University Press.

Arrhenius, S. (1896) 'On the influence of carbonic acid in the air upon the temperature of the round', *London, Edinburgh, and Dublin Philosophical Magazine and Journal of Science*, 41: 237–275.

Arrhenius, S. (1908) *Worlds in the Making,* New York: Harper & Brothers.

Auge, M. (1995) *Non-Places: Introduction to an Anthropology of Super-modernity*, London: Verso.

Bailey, A. (2005) *Making Population Geography*, London: Hodder Arnold.

Baran, S. and Wallis, R. (1990) *The Known World of Broadcast News: International News and the Electronic Media*, London: Routledge.

Barber, B. (1996) *Jihad vs McWorld: How Globalism and Tribalism are Reshaping the World*, New York: Ballantine Books.

Barkan, E. (1992) *The Retreat from Scientific Racism*, Cambridge: Cambridge University Press.

Barnett, T. et al., (2005) 'Penetration of human-induced warming into the world's oceans', *Science*, 309: 284–287.

Bastian, O. (2002) 'Landscape ecology – towards a unified discipline', *Land-scape Ecology*, 16: 757–766.

Benko, G. and Strohmayer, U. (2004) 'Conclusion, or an introduction to human geography in the 21st century', in Georges Benko and Ulf Strohmayer (Eds) *Human Geography: A History for the 21st Century*, London: Arnold.

Berghoff, H., Harvie, C., Korte, B. and Schneider, R. (Eds) (2002) *The Making of Modern Tourism: The Cultural History of the British Experience, 1600–2000*, London: Palgrave.

Bessette, G. (2004) *Involving the Community: A Guide to Participatory Development Communication*, Ottawa: International Development Research Centre.

Black, J. (2003) *Visions of the World: A History of Maps*, London: Mitchell Beazley.

Bonnett, A. (2003) 'Geography as the world discipline: connecting popular and academic geographical imaginations', *Area*, 35 (1): 55–63.

Bonnett, A. (2005) 'Occidentalism and plural modernities: or how Fukuzawa and Tagore invented the West', *Environment and Planning D: Society and Space*, 23: 505–525.

Boserup, E. (1980) *Population and Technology*, Oxford: Blackwell.

Bowen, M. (1981) *Empiricism and Geographical Thought: From Francis Bacon to Alexander von Humbolt*, Cambridge: Cambridge University Press.

Bowler, P. (1992) *The Fontana History of the Environmental Sciences*, London: Fontana.

Bowman, I. (1928) *The New World: Problems in Political Geography: Fourth Edition*, Yonkers-on-Hudson, New York: World Book Company.

Bowman, I. (1934) *Geography in Relation to the Social Sciences*, New York: Charles Scribner's Sons.

Brunhes, J. (1920) *Human Geography*, Chicago, IL: Rand McNally.

Brunn, S. and Williams, J. (1983) *Cities of the World: World Regional Urban Development*, New York: Harper and Row.

Bryson, J. and Henry, N. (2001) 'The global production system: from Fordism to post-Fordism', in P. Daniels et al. (Eds) *Human Geography: Issues for the 21st Century*, London: Prentice Hall.

Bunge, W. (1971) *Fitzgerald: Geography of a Revolution*, Cambridge: Schenkman Publishing.

Bunnell, T., Drummond, L. and Chong, H. (Eds) (2002) *Critical Perspectives on Cities in Southeast Asia,* Singapore: Times Academic Press.

Burgess, E. (1925) 'The growth of the city: an introduction to a research project', in R. Park, E. Burgess and R. McKenzie (Eds) *The City*, Chicago: University of Chicago Press.

Burkholder, M. and Johnson, L. (1994) *Colonial Latin America*, New York: Oxford University Press.

Carr, E. H. (1987) *What is History?*, London: Penguin.

Carson, R. (1962) *Silent Spring*, Boston, MA: Houghton Mifflin.

Castree, N. (2005) 'Whose geography? Education as politics', in N. Castree, A. Rogers and D. Sherman (Eds) *Questioning Geography*, Oxford: Blackwell.

Castree, N. (Ed.) (2006) *David Harvey: A Critical Reader, Oxford:* Blackwell.

Chatwin, B. (1977) *In Patagonia*, London: Jonathan Cape.

China Defence (2006) Delphi expects $1 billion China sales', http://www.china-defense.com/forum/showthread.php?t=8620&page=6, accessed 27.12.2006.

Clifford, J. (1988) *The Predicament of Culture: Twentieth-Century Ethnography, Literature, and Art*, Cambridge, MA: Harvard University Press.

Cloke, P. and Little, J. (Eds) (1997) *Contested Countryside Cultures*, London: Routledge.

Commoner, B. (1971) *The Closing Circle: Nature, Man and Technology*, New York: Knopf.

Commoner, B. (1990) 'Fundamental causes of the environmental crisis', in Roderick Nash (Ed.) *American Environmentalism: Readings in Conservation History: Third Edition*, New York: McGraw-Hill.

Comparative Risk Assessment Collaborating Group (2002) 'Selected major risk factors and global and regional burden of disease', *Lancet*, 360: 1347–1360.

Conrad, J. (1926) *Last Essays*, London: J.M. Dent.

Cosgrove, D. (2001) *Apollo's Eye: A Cartographic Genealogy of the Earth in the Western Imagination*, Baltimore, MA: Johns Hopkins University Press.

Cotgrove S. and Duff, A. (1980) 'Environmentalism, middle-class radicalism and politics', *Sociological Review*, 28 (2): 333–51.

Cresswell, T. (2004) *Place: A Short Introduction*, London: Blackwell.

Crosby, A. (1986) *Ecological Imperialism: The Biological Expansion of Europe, 900–1900*, Cambridge: Cambridge University Press.

Dansgaard, W., Johnsen, S. et al. (1993) 'Evidence for general instability of climate from a 250–Kyr ice-Core record', *Nature*, 364: 218–220.

Davey, S. (2004) *Unforgettable Places to See Before you Die*, London: BBC Books.

Davis, P. (1984) *The Flora of Turkey and the East Aegean Islands: Volume One*, Edinburgh: Edinburgh University Press.

Dear, M. (2000) *The Postmodern Urban Condition*, Oxford: Blackwell.

Dear, M. and Flusty, S. (2002) 'Introduction', in M. Dear and S. Flusty (Eds) *The Spaces of Postmodernity*, Oxford: Blackwell.

de Blij, H. (2005) *Why Geography Matters: Three Challenges Facing America: Climate Change, the Rise of China, and Global Terrorism*, Oxford: Oxford University Press.

Debord, G. (1983) *Society of the Spectacle*, Detroit: Black and Red.

Debord, G. (1991) *Panegyric*, London: Verso.

Delanty G. (Ed.), (2006) *Europe and Asia Beyond East and West*, London: Routledge.

de la Fuente Layos, L. (2005) *Statistics in Focus: Transport: Short Distance Passenger Mobility in Europe*, Luxembourg: Eurostat.

de Montaigne, M. (1993) *The Complete Essays*, London: Penguin.

Diamond, J. (1998) *Guns, Germs and Steel: A Short History of Everybody for the Last 13,000 Years*, London: Vintage.

Diamond, J. (2005) *Collapse: How Societies Choose to Fail or Succeed*, London: Allen Lane.

Dikotter, F. (1992) *The Discourse of Race in Modern China*, Stanford: Stanford University Press.

Dirlik, A. (2001) 'Place-based imagination: globalism and the politics of place', in R. Prazniak and A. Dirlik (Eds) *Places and Politics in an Age of Globalization*, Lanham: Rowman and Littlefield.

Donald, J. (1997) 'This, here, now: imagining the modern city', in S. Westwood and J. Williams (Eds) *Imagining Cities: Scripts, Signs, Memories*, London: Routledge.

Earth Trends, (2006) 'Population, Health and Human Well-being – urban and rural areas: urban population as a percentage of total population', http://earthtrends.wri.org/text/population-health/variable-448.html, accessed 22.12.2006.

East Asian History Sourcebook, *Chinese Accounts of Rome, Byzantium and the Middle East, c. 91 B.C.E.-1643 C.E.*, http://depts.washington.edu/silkroad/texts/romchin1.html, accessed 20.12.2006.

Ehrlich, P. and Holdren, J. (1971) 'Impact of population growth', *Science*, 171: 1212–1217.

Eisenstadt, S. (2000) 'Multiple modernities', *Daedalus*, 129 (1): 1–29.

Ekholm, N. (1901) 'On the variations of the climate of the geological and Historical Past and Their Causes', *Quarterly Journal of Royal Meteorological Society*, 27: 1–61.

Febvre, L. (1925) *A Geographical Introduction to History*, London: Knopf.

Fenneman, N. (1919) 'The circumference of geography', *Annals of the Association of American Geographers*, 9: 3–11.

Fisher, J. (1998) *Nongovernments: NGOs and the Political Redevelopment of the Third World*, Bloomfield, CT: Kumarian Press.

Frobel, F., Heinrichs, J. and Kreye, O. (1980) *The New International Division of Labour*, Cambridge: Cambridge University Press.

Fukuyama, F. (1992) *The End of History and the Last Man*, London: Hamish Hamilton.

Fukuzawa, Y. (1869) *Sekai Kunizukushi [World Geography]*, Tokyo: Okadaya Kashichi.

Fukuzawa, Y. (1973) *An Outline of a Theory of Civilization*, Tokyo: Sophia University.

Gill, G. (1903) *Gill's Student's Geography: Sixth Edition*, London: George Gill and Sons.

Glacken, C. (1976) *Traces on the Rhodian Shore: Nature and Culture in Western Thought from Ancient Times to the End of the Eighteenth Century*, Berkeley: University of California Press.

Geertz, C. (1988) *Works and Lives: The Anthropologist as Author*, Stanford: Stanford University Press.

General Motors (2006) http://www.gm.com/company/corp_info/history/gmhis1920.html, accessed 26.12.2006.

Gottdiener, M. and Budd, L. (2005) *Key Concepts in Urban Studies*, London: Sage.

Gottmann, J. (1961) *Megaopolis: The Urbanized North-Eastern Seaboard of the United States*, Cambridge, MA: MIT Press.

Goudie, A. and Viles, H. (1997) *The Earth Transformed: An Introduction to Human Impacts on the Environment*, Oxford: Blackwell.

Gould, P. and White, R. (1985) *Mental Maps*, London: Routledge.

Green, J. (1879) 'Introduction', in J.R. Green and A. Green, *A Short Geography of the British Isles*, London: Macmillan.

Gregory, D. (1978) *Ideology, Science and Human Geography*, London: Hutchinson.

Gregson, N. (2003) 'Discipline games, disciplinary games and the need for a post-disciplinary pratice: responses to Nigel Thrift's "The future of geography"', *Geoforum*, 34: 5–7.

Grenier, L. (1998) *Working with Indigenous Knowledge: A Guide for Researchers*, Ottawa: International Development Research Centre.

Grimmett, R. (1999) *Birds of India: Pakistan, Nepal, Bangladesh, Bhutan, Sri Lanka and the Maldives*, Princeton: Princeton University Press.

Gwynne, R., Klak, T. and Shaw, D. (2003) *Alternative Capitalisms: Geographies of Emerging Regions*, London: Arnold.

Habhan, G. and Trimble, S. (1994) *The Geography of Childhood: Why Children Need Wild Places*, Boston, MA: Beacon Press.

Haggett, P. (1995) *The Geographer's Art*, Oxford: Blackwell.

Hagman, O. (2006) 'Morning queues and parking problems: on the broken promises of the automobile', *Mobilities*, 1 (1): 63–74.

Hammer, T. (1990) *Democracy and the Nation State: Aliens, Denizens and Citizens in a World of International Migration*, Aldershot: Avebury.

Handwerk, B. (2006) 'China's car boom tests safety, pollution practices', http://geography.about.com/gi/dynamic/offsite.htm?site=http://news.national geographic.com/news/2004/06/0628%5F040628%5Fchinacars.html, accessed 27.12.2006.

Harootunian, H. (2000) *Overcome by Modernity: History, Culture and Community in Interwar Japan*, Princeton: Princeton University Press.

Harris, P. (2004) '"Getting rich is glorious": environmental values in the People's Republic of China', *Environmental Values* 13 (2): 145–165.

Hartshorne, R. (1959) *Perspectives on the Nature of Geography*, Chicago, IL: Rand McNally.

Hartshorne, R. (1961) *The Nature of Geography: A Critical Survey of Current Thought in the Light of the Past*, Lancaster: Association of American Geographers.

Harvey, D. (1977) 'Geography policies, financial institutions and neighbourhood change in United States cities, in M. Harloe (Ed.) *Captive Cities*, London: John Wiley.

Harvey, D. (1978) 'The urban process under capitalism: a framework for analysis', *International Journal of Urban and Regional Research*, 2: 101–31.

Harvey, D. (1982) *The Limits to Capital*, Oxford: Basil Blackwell.

Harvey, D. (1990) *The Condition of Postmodernity. An Enquiry into the Origins of Cultural Change*, Oxford: Blackwell.

Harvey, G. (2006) *We Want Real Food*, London: Constable.

Her Majesty's Treasury (2006) *The Stern Review on the Economics of Climate Change*, London: HM Treasury.

Hanson, V. (2001) *Why the West Won: Carnage and Culture from Salamis to Vietnam*, London: Faber and Faber.

Hay, S. (1970) *Asian Ideas of East and West: Tagore and his Critics in Japan, China, and India*, Cambridge, MA: Harvard University Press.

Hayden, B. (1981) 'Subsistence and ecological adaptations of modern hunter-gatherers', in R. Harding and G. Teleki (Eds) *Omnivorous Primates*, New York: Columbia University Press.

Heartfield, J. (2004) 'Londonostalgia', reprinted from *Blueprint*, September 2004, http://www.design4design.com/artucles/artcles_story.asp? STO RYID=5765, accessed 05/04/2005.

Hegel, G. (1991) *The Philosophy of History*, Amherst: Prometheus Books.

Hettner, A. (1927) *Die Geographie, ihre Geschichte, ihr Wesen und ihre Methoden*, Breslau: Ferinand Hirt.

Hispanic CREO, National Latino Statistics, (2006) http://www.justiceby choice.org/National%20Latino%20Statistics.pdf, accessed 26.12.2006.

Holdren, J. and Ehrlich, P. (1974) 'Human population and the global environment, *American Scientist*, 62 (3): 282–292.

Howard, E. (1902) *Garden Cities of To-morrow*, London: Swan Sonnenschein.

Hu, H. (1947) [1942] *Shijie Dili [Geography of the World]*, Shanghai: Zhengzhong Shuju.

Hutton, J. (1788) 'Theory of the Earth: an INVESTIGATION into the Laws observable in the Composition, Dissolution, and Restoration of Land upon the Globe', *Transactions of the Royal Society of Edinburgh*, 1, 2, 209–304.

Intergovernmental Panel on Climate Change (2001) *Climate Change 2001: Summary for Policy Makers,* Cambridge: Cambridge University Press.

International Panel on Climate Change (2007) *Climate Change 2007: Impacts, Adaptation and Vulnerability: Working Group II Contribution to the Intergovernmental Panel on Climate Change Fourth Assessment Report: Summary for Policymakers*, Geneva: IPCC.

Jacobs, J. (1961) *The Death and Life of Great American Cities*, New York: Random House.

Jameelah, M. (1979) *Western Civilization Condemned by Itself: A Comprehensive Study of Moral Retrogression and its Consequences. Volume 1*, Lahore: Mohammad Yusaf Khan and Sons.

Johnson, J. (1967) *Urban Geography: An Introductory Analysis*, Oxford: Pergamon Press.

Kejariwal, O. (1999) *The Asiatic Society of Bengal and the Discovery of India's Past*, New Delhi: Oxford University Press.

Kidd, B. (1894) *Social Evolution*, London: Macmillan.

Kierman, B. (1996) *The Pol Pot Regime: Race, Power, and Genocide in Cambodia under the Khmer Rouge, 1975–79*, New Haven, CT: Yale University Press.

Kimble, G. (1951) 'The inadequacy of the regional concept', in L. Stamp and S. Wooldridge (Eds) *London Essays in Geography*, London: Longmans, Green.

Kingsnorth, P. (2004) *One No, Many Yeses: A Journey to the Heart of the Global Resistance Movement*, London: Free Press.

Kipling, R. (1939) *Sixty Poems*, London: Hodder and Stoughton.

Kitchin, R. and Tate, N. (2000) *Conducting Research into Human Geography*, Harlow: Prentice Hall.

Kropotkin, P. (1996) 'What geography ought to be', in J. Agnew, D. Livingstone and A. Rogers (Eds) *Human Geography: An Essential Anthology*, Oxford: Blackwell.

Kunstler, J. (1993) *The Geography of Nowhere*, New York: Simon and Schuster.

Lamprey, H. (1981) 'Pastoralism yesterday and today: the over-grazing problem', in J. Bourliere (Ed.) *Tropical Savannas: Ecosystems of the World, Volume 13*, London: Elsevier.

Latouche, L. (1996) *The Westernization of the World: Significance, Scope and Limits of the Drive Towards Global Uniformity*, Oxford. Polity Press.

Lees, L. (2003) 'Urban geography: The "death" of the city?', in A. Rogers and H. Viles (Eds) *The Student's Companion to Geography: Second Edition*, London: Blackwell.

Leonard, J. (1984) *Wei Yuan and China's Rediscovery of the Maritime World*, Cambridge: Harvard University Press.

Leopold, A. (1966[1949]) *Sand County Almanac: with other Essays on Conservation from Round River*, New York: Oxford University Press.

Livingstone, D. (1992) *The Geographical Tradition: Episodes in the History of a Contested Enterprise*, Oxford: Blackwell.

Lomborg, B. *The Sceptical Environmentalist*, Cambridge: Cambridge University Press, Cambridge.

Lorenz, E. (1966) *Nonlinearity, Weather Prediction, and Climate Deduction*, Cambridge: Massachusetts Institute of Technology, Deptartment of Meteorology.

Lorenz, E. (1996) *The Essence of Chaos*, Seattle: University of Washington Press.

Lovejoy, P. (2000) *Transformations in Slavery*, Cambridge: Cambridge University Press.

Lovelock, J. (2000) *Gaia: A New Look at Life on Earth*, Oxford: Oxford Paperbacks.

Lowell, B. (2006) 'Immigrants and labor force trends: the future, past, and present', http://www.migrationpolicy.org/ITFIAF/TF17_Lowell.pdf, accessed 26.12.2006.

MacCannell, D. (1992) *Empty Meeting Grounds: The Tourist Papers*, London: Routledge.

Mace, G., Masundire, H. and Baillie, J. (2005) 'Biodiversity', in R. Hassan, R. Scholes and N. Ash (Eds) *Ecosystems and Human Well-Being: Volume 1*, Washington: Island Press.

Mackinder, H. (1996) 'On the scope and methods of geography', in J. Agnew, D. Livingstone and A. Rogers (Eds) *Human Geography: An Essential Anthology*, Oxford: Blackwell.

Malinowski, B. (1922) *Argonauts of the Western Pacific*, London: George Routledge and Sons.

Malone, K. and Tranter, P. (2003) 'School grounds as sites of learning: making the most of environmental opportunities', *Environmental Education Research*, 9 (3): 283–303.

Malthus, T. (1976/1978) *An Essay on the Principle of Population*, Harmondsworth: Pelican.

Mandelbaum, M. (2002) *The Ideas that Conquered the World: Peace, Democracy and Free Markets in the Twenty-First Century*, New York: Public Affairs.

Marsh, G. (1965[1864]) *Man and Nature*, Cambridge: Harvard University Press.

Martin, G. and James, P. (1993) *All Possible Worlds: A History of Geographical Ideas*, London: John Wiley.

Martin, T. (1998) 'The origins of Soviet ethnic cleansing', *Journal of Modern History*, 70: 813–861.

Marx, K. (1973) *The Revolutions of 1848*, Harmondsworth: Penguin.

Matthews, J. and Herbert, D. (2004) *Unifying Geography: Common Heritage, Shared Future*, London: Routledge.

Massey, D. (1997) 'A global sense of place', in T. Barns and D. Gregory (Eds) *Reading Human Geography*, London: Arnold.

Massey, D. (2005) *For Space*, London: Sage.

Massey, D. (2007) *World City,* Cambridge: Polity Press.

Mattingly, D. (2001) 'The home and the world: domestic service and international networks of caring labor', *Annals of the Association of American Geographers*, 91: 370–86.

May, J. (1970) *Kant's Concept of Geography and its Relation to Recent Geographical Thought*, Toronto: University of Toronto Press.

McDowell, L. (Ed.) (1997) *Undoing Place? A Geographical Reader*, London: Arnold.

McKibben, B. (2003) *The End of Nature: Humanity, Climate Change and the Natural World*, London: Bloomsbury.

Meadows, D., Meadows, D., Randers, J. and Behrens, W. (1972) *Limits to Growth: A Report For The Club Of Rome's Project On The Predicament Of Mankind*, New York: Potomac Associates.

Meentemeyer, V. (1989) 'Geographical perspectives of space, time, and scale', *Landscape Ecology*, 3 (3–4): 163–173.

Meyer, W. and Turner, B. (1995) 'The Earth transformed: trends, trajectories, and patterns', in R. Johnston, P. Taylor and M. Watts (Eds) *Geographies of Global Change*, Oxford: Blackwell.

Milankovitch, M. (1920) *Theorie Mathematique des Phenomenes Thermiques produits par la Radiation Solaire*, Paris: Gauthier-Villars.

Milankovitch, M. (1930) *Mathematische Klimalehre und Astronomische Theorie der Klimaschwankungen, Handbuch der Klimalogie Band 1*, Berlin: Teil A Borntrager.

Mitchell, W. (1995) *City of Bits: Space, Place and the Infobahn*, Cambridge, MA: MIT Press.

Miyoshi, M. and Harootunian, H. (2002) 'Introduction: the 'afterlife' of Area Studies', in M. Miyoshi and H. Harootunian (Eds) *Learning Places: the Afterlives of Area Studies*, Durham: Duke University Press.

Montesquiew (1973) *Persian Letters*, London: Penguin.

Morris, M. (1992) 'The Man in the Mirror: David Harvey's "Condition" of Postmodernity', *Theory, Culture, Society*, 9: 253–279.

Morris, W. (1994) *Political Writings: Contributions to Justice and Commonweal 1883–1890*, Bristol: Thoemmes Press.

Morse, J. (1784) *Geography Made Easy*, New Haven, CT: Meigs, Bowen and Dana.

Mumford, L. (1966) *The City in History*, Harmondsworth: Penguin.

Murdoch, J., Lowe, P., Ward, N. and Marsden, T. (2003) *The Differentiated Countryside*, London: Routledge.

Murray, W. (2006) *Geographies of Globalization*, London: Routledge.

Myers, N. (2005) 'Environmental Refugees: An Emergent Security Issue', paper delivered at 13th Economic Forum, Prague, 23–27 May.

Nast, H. (1994) 'Opening remarks on "women in the field"', *Professional Geographer*, 46: 54–66.

Nelson, A. (2006) 'By 2030 the US will have re-built almost half its built environment', http://www.citymayors.com/development/built_environment_usa.html, accessed 22.12.2006.

Nepstad, D. et al. (2007) *The Amazon in a Changing Climate: Large-Scale Reductions of Carbon Emissions from Deforestation and Forest Impoverishment*, http://www.whrc.org/resources/published_literature/pdf/Amazon-and-Climate-2006.pdf, accessed 8.4.2007.

News from Nowhere (Ed.) (2003) *We Are Everywhere: The Irresistible Rise of Global Anti-Capitalism*, London: Verso.

Nietz, J. (1961) *Old Textbooks: Spelling, Grammar, Reading, Arithmetic, Geography, American History, Civil Government, Physiology, Penmanship, Art, Music, as Taught in the Common Schools from Colonial Days to 1900,* Pittsburgh: University of Pittsburgh Press.

Paterson, J. (1974) 'Writing regional geography: problems and progress in the Anglo-American realm', *Progress in Human Geography*, 6. 1–16.

People's Daily Online, 'China encourages mass urban migration', http:// english.people.com.cn/200311/28/eng20031128_129252.shtml, accessed 22.12.2006.

Pernia E. (Ed.) (1999) *Urban Poverty in Asia: A Survey of Critical Issues,* Hong Kong: OUP China.

Pliny the Elder (1991) *Natural History: A Selection,* London: Penguin.

Ploszajska, T. (1999) *Geographical Education, Empire and Citizenship: Geographical Teaching and Learning in English Schools, 1870–1944,* London: Historical Geography Research Group.

Polyan, P. (2001) *Не по своей воле...История и география принудительных миграций в СССР,* Moscow: ОГИ Мемориал .

Porter, B. (1968) *Critics of Empire: British Radical Attitudes to Colonialism in Africa 1895–1914,* London: Macmillan.

Ravenstein, G. (1889) 'The laws of migration', *Journal of the Royal Statistical Society*, 52 (2): 241–305.

Read, D. (1999) *The Power of News: The History of Reuters: Second Edition,* Oxford: Oxford University Press.

Relph, E. *Place and Placelessness,* London: Pion.

Rex, J. (1968) 'The sociology of a zone of transition', in Ray Pahl (Ed.) *Readings in Urban Sociology,* London: Pergamon.

Ritter, C. (1865) *Comparative Geography,* Edinburgh: William Blackwood and Sons.

Roberts, J. (1985) *The Triumph of the West,* London: BBC.

Robinson, G. (1998) *Techniques and Methods in Human Geography,* London: Wiley.

Romm, J. (1992) *The Edges of the Earth in Ancient Thought: Geography, Exploration and Fiction,* Princeton: Princeton University Press.

Rose, G. (1993) *Feminism and Geography,* Cambridge: Polity Press.

Rosemont, F. (1988) Surrealists on whiteness from 1925 to the present', *Race Traitor*, 9: 5–18.

Rousseau, J-J, (1984 [1755]) *A Discourse on Inequality,* London: Penguin.

Rystad, G. (1992) 'Immigration history and the future of international migration', *International Migration Review*, 26(4): 1168–1199.

Sale, K. (1990) 'Schism in environmentalism', Roderick Nash (Ed.) *American Environmentalism: Readings in Conservation History; Third Edition,* New York: McGraw-Hill.

Said, E. (1978) *Orientalism: Western Representations of the Orient,* Harmondsworth: Penguin.

Sample, I. (2006) 'Forests are poised to make a comeback, study shows', *The Guardian*, November 14.

Sassen, S. (2001) *The Global City: London, New York, Tokyo*, Princeton: Princeton University Press.

Saunders, P. (1981) *Social Theory and the Urban Question*, London: Hutchinson.

Scottish Executive (2006) 'Statistical Bulletin Transport Series: Trn/2005/3: Travel by Scottish residents: some National Travel Survey results for 2002/2003 and earlier years', http://www.scotland.gov.uk/Publications/2005/04/1894658/46593, accessed 28.12.2006.

Seneca, L. (1973) *Medea*, Oxford: Oxford University Press.

Sennett, R. (1994) *Flesh and Stone: The Body and the City in Western Civilization*, London: Faber and Faber.

Schneider S. and Boston, P. (Eds) (1991) *Scientists on Gaia*, Cambridge, MA: MIT Press.

Schultz, P. (2003) *1000 Places to See Before you Die*, New York: Workman Publishing.

Schumacher, E. (1973) *Small Is Beautiful. A Study Of Economics As If People Mattered*, London: Blond and Briggs.

Siegenthaler U., Stocker, T. et al. (2005) 'Stable carbon cycle-climate relationship during the late Pleistocene', *Science*, 310: 1313–1317.

Simmel, G. (1997) *Simmel on Culture: Selected Writings*, London: Sage.

Sinclair, I. (2003) *London Orbital*, London: Penguin.

Sinclair, I. (2006) *London: City of Disappearances*, London: Hamish Hamilton.

Smil, V. (2001) *Enriching the Earth*, Cambridge, MA: MIT Press.

Smith, N. (1982) '"Academic wars over the field of geography": the elimination of geography at Harvard, 1947–1952', *Annals of the Association of American Geographers*, 77: 155–72.

Smith, P. (1998) *The History of Tourism: Thomas Cook and the Orgins of Leisure Travel*, London: Routledge.

Spencer, J. (1995) 'Occidentalism in the East: the uses of the West in the politics and anthropology of South Asia', in J. Carrier (Ed.) *Occidentalism: Images of the West*, Oxford: Oxford University Press.

Stamp, D. (1948) *The Land of Britain: Its Use and Misuse*, London: Longman.

Stanley, H. M. (1885) 'Central Africa and the Congo Basin; or, the importance of the scientific study of geography', *Journal of the Manchester Geographical Society*, 1: 1–3, 6–25.

Stevens, S. (2001) 'Fieldwork as commitment', *The Geographical Review*, 91: 66–73.

Stoddart, D. (1986) *On Geography*, Oxford: Blackwell.

Strabo, 'The *Geography* of Strabo', http://penelope.uchicago.edu/Thayer/E/Roman/ Texts/Strabo/home.html, accessed 20/12/2006.

Suwastoyo, B. (2006) 'Haze From Indonesia Fires Chokes Region, Spreads Across Pacific', http://www.terradaily.com/reports/Haze_From_Indonesia_Fires_Chokes_Region_Spreads_Across_Pacific_999.html, accessed 21.12.2006.

Swift, J. (1733) *On Poetry, a Rapsody*, Dublin and London: Robert Fleming.

Tagore, R. (1922) *Creative Unity*, London: Macmillan.

Tatham, G. (1957) 'Environmentalism and possibilism', in G. Taylor (Ed.) *Geography in the Twentieth Century: A Study of Growth, Fields, Techniques, Aims and Trends* (Third Edition), London: Methuen.

Tobler, W. (1995) 'Migration: Ravenstein, Thorntwaite, and beyond', *Urban Geography*, 16(4): 327–343.

Tönnies, T. (1963) *Community and Society*, New York: Harper and Row.

Turner, S., Hanham R. and Portararo, A. (1977) 'Population pressure and agricultural intensity', *Annals of the Association of American Geographers*, 67: 386–97.

United Nations, Department of Economic and Social Affairs, (2006) 'Population Division World Urbanization Prospects: The 2005 Revision', http://www.un.org/esa/population/publications/WUP2005/2005wup.htm, accessed 22.12.2006.

United Nations News Center (2006) 'World population to reach 9.1 billion in 2050, UN projects', http://www.un.org/apps/news/story.asp? NewsID=13451&Cr=population&Cr1, accessed 22.12.2006.

Vidal de la Blache, P. (1922) *Principes de geographie humaine*, Colin, Paris, 1922.

Walter, K., Zimcu, S. et al. (2006) 'Methane bubbling from Siberian thaw lakes as a positive feedback to climate warming', *Nature*, 443: 71–75.

Watkins, S. and Jones, C. (2000) *Unforgettable Journeys to Take Before You Die*, London: BBC Books.

Weart, S. (2003) *The Discovery of Global Warming*, Cambridge, MA: Harvard University Press.

Wei-ming, T. (2001) 'The ecological turn in new Confucian humanism: implications for China and the world', *Daedalus*, 130 (4): 243–264.

Wegener, A. (1966) *The Origin of Continents and Oceans*, New York: Dover Publications.

Wei, Y. (2002), 'Multiscale and multimechanisms of regional inequality in China: implications for regional policy', *Journal of Contemporary China*, 11 (30): 109–124.

Wickham, T. (1993) 'Farmers ain't no fools: exploring the role of participatory rural appraisal to access indigenous knowledge and enhance sustainable development research and planning. A case study of Dusun Pausan, Bali, Indonesia', Master's Thesis, Faculty of Environmental Studies, University of Waterloo.

Wirth, L. (1938) 'Urbanism as a way of life', *American Journal of Sociology*, 44: 1–24.

Wooldridge, S. and East, W. (1951) *The Spirit and Purpose of Geography*, New York: Hutchinson's University Library.

Wordsworth, W. (1994) 'Lines Written a Few miles above Tintern Abbey, on Revisiting the Banks of the Wye During a Tour, 13 July 1798', in Duncan Wu (Ed.) *Romanticism: An Anthology*, Oxford: Blackwell.

World Tourism Organisation (2006) The Chinese Outbound Tourism Market, http://www.world-tourism.org/newsroom/Releases/2006/november/chinese outbound.htm, accessed 27.12.2006.

Wright, J. (1965) *The Geographical Lore of the Time of the Crusades: A Study in the History of Medieval Science and Tradition in Western Europe*, New York: Dover Publications.

Xu, J. (1986) [1848] *Yihghuan Zhilue [Brief Account of the Maritime Circuit]*, Tabei, Taiwan: Wenhai Chubanshe.

Yeats, W. (1913) 'Introduction', in R. Tagore, *Gitanjali (Song Offerings)*, London: Macmillan.

Index

Pan Gu 9
Paradise 63
patriotism 27
Patterson, J. 109
peasants 21, 64–5
Persian Letters 84
*Perspectives on the Nature of
    Geography* 110
Peters, A. 93
The Philosophy of History 15
Physiography 82
Place and Placelessness 67
place (decline of) 21, 66–9, 79, 133
    n.41, 133 n.42
plate tectonics 51
play 99–100
Pliny the Elder 18, 129 n.58
Ploszajska, T. 82
Political Geography 108
pollution 30, 36, 38–47, 91, 122
Pol Pot 65
Portararo, A. 46
popular geography 4–5, 27, 114–21
population growth (human) 30, 39,
    45–7, 122
possiblism 33
post-disciplinarity 112
post-modernism 62, 67–9
primitivism 19–20, 26

'race' 91
Ratzel F. 108
Ravenstein, E. 73
Ray, J. 10–11
refugees 43
regional geography 11, 34,
    90, 109–110
religion 9–14, 16–17, 35–6, 63–4
Relph, E. 67
Reuters 117–8
Rex, J. 61
Rio Declaration 106
Ritter, C. 87, 107, 109
Roberts, J. 12
Roman empire 8, 17, 19, 20, 57
Rose, G. 82
Rousseau, J-J. 26

Royal Geographical Society 24, 81, 86,
    102–3, 107, 114
rural life 62, 65–7
rural population 39–40, 55, 70
Russia 15–16, 20
Rystad, G. 70

Said, E. 15
Salzman, L. 116
Sand County Almanac 36
satellites 30, 94, 118
Seneca, L. 8
Severian of Gabala 14
Schneider, S. 53
school geography 5, 23, 26–7, 82, 102,
    104–5
Scotland 77
Scott, R. 85–6
Scythians 19–20
Sennett, R. 63–4
Silent Spring 38
*A Short Geography of the
    British Isles* 105
Simmel, G. 63, 69
Sinclair, I. 87
Sites of Special Scientific Interest 83
situationism 64
slavery 70–1
Sloan, A. 76
Small is Beautiful 38
Smil, V. 45
sociology 28, 38, 59
soil 30, 38, 44–5, 96
Solidar 116
Stalin, J. 71
Stanley, H.M. 21–2
Steers, A. 82–3
The Stern Review 42
Stockholm Declaration 106
Stoddard, D. 123–4, n.3
Stonehenge 14
Strabo 18–20, 32, 56–7
space, concept in human geography
    111–12
space travel 2
spatial analysis 111
species diversity 36, 38–9, 43–7